The Old Barn Book

THE
OLD BARN
BOOK

A
Field Guide to
North American Barns
and Other Farm Structures

Allen G. Noble
A N D
Richard K. Cleek

with Illustrations by
M. Margaret Geib

Rutgers University Press

NEW BRUNSWICK, NEW JERSEY

Sixth paperback printing, 2007

Library of Congress Cataloging-in-Publication Data

Noble, Allen George, 1930–
    The old barn book : a field guide to North American barns and other farm
structures / Allen G. Noble and Richard K. Cleek ; with illustrations by M.
Margaret Geib.
        p.   cm.
    Based on v. 2, Barns and farm structures, of A. G. Noble's Wood, brick, and
stone, published in 1984 by the University of Massachusetts Press, Amherst.
    Includes bibliographical references (p.   ) and index.
    ISBN 0-8135-2172-6 (cloth) — ISBN 0-8135-2173-4 (pbk.)
    1. Barns—United States.   2. Barns—Canada.   3. Farm buildings—United
States.   4. Farm buildings—Canada.   5. Outbuildings—United States.   6. Outbuildings—Canada.   7. Fences—United States.   8. Fences—Canada.   I. Cleek,
Richard K., 1945–   . II. Geib, M. Margaret.   III. Noble, Allen George,
1930–   . Wood, brick, and stone. 2. Barns and farm structures.   IV. Title.
NA8230.N63   1995
728'.922'0973—dc20   94-41300
                                                                            CIP

British Cataloging-in-Publication information available

Contents

STRAW SHEDS AND HORSEPOWER SHEDS •
BARN DECORATION •
Door Decoration • Barn Murals • Decorative Roofs •

ENGLISH BARNS •
*Three-Bay Threshing Barn • New England Connected
Barn • English Bank Barn • Raised Barn • Foundation
Barn • Welsh Gable-Entry Bank Barn •*

GERMAN BARNS •
*Grundscheier • Grundscheier with Forebay •
Grundscheier Type D • Classic Sweitzer Barn •
Standard Pennsylvania Barn • Special Forms of the
Standard Pennsylvania Barn • Extended Pennsylvania
Barn • Double-Decker Barn • Madison County Amish
Barn •*

FRENCH BARNS •
*Quebec Long Barn • Madawaska Twin Barn • Acadian
Barn • Cajun Barn •*

HISPANIC BARNS •
Tasolera • Tapiesta • Hispanic Twin-Crib Barn •

OTHER ETHNIC BARNS •
*Dutch Barn • Swedish Barn • Finnish Barns • Czech
Barn •*

OTHER REGIONAL BARNS •
*Manitoba Mennonite Housebarn • Mormon Thatched-
Roof Cowshed •*

*Three-End Barn • Feeder Barn • Erie Shore Barn •
Round-Roof Barns • Pole Barn • Polygonal and
Round Barns •*

Preface

The Old Barn Book: A Field Guide to North American Barns and Other Farm Structures aims to provide a convenient and inexpensive handbook with which one can identify the thousands of structures that dot the rural landscape of North America. Its descriptions, drawings, and photos give essential information about these fascinating and fast-disappearing structures, in a format handy enough to be taken along on a drive around any part of rural North America.

Today most Americans and Canadians are born in cities, live much of their lives in urban environments, and have little experience of farmsteads and countrysides. But interest in farm buildings has not declined. The environmental conservation movement, combined with architectural preservation efforts, historically oriented celebrations such as the U.S. Bicentennial, and the success of open-air museums such as Upper Canada Village and Old World Wisconsin have increased the appeal of rural areas and their buildings. People from all walks of life and all kinds of backgrounds yearn to know more about their collective roots. We hope this handbook will help unlock some of the secrets of the countryside and at the same time provide insight into what gives cultural character to the different regions of the continent.

The field guide is based on a larger work by Allen G. Noble titled *Wood, Brick, and Stone: The North American Settlement Landscape* published by the University of Massachusetts Press. The second volume of that work, *Barns and Other Farm Structures,* provides a more detailed discussion and more references than are given here.

We hope this manual gives you many pleasant hours as you search out the barns, silos, fences, and other farm features. Happy barn hunting!

Allen G. Noble
Richard K. Cleek

Acknowledgments

A field guide to aid interested readers in identifying the rural farm structures of North America was first suggested by Richard Cleek, who had read Allen Noble's earlier work, *Wood, Brick, and Stone*. Together we decided to assemble a volume patterned after the field guides that have become so popular and useful in the natural sciences. The process has been long and complicated but, on balance, most enjoyable.

Several people have aided us with their expertise. All or part of the manuscript was read by Greg Huber, Bob Ensminger, Warren Roberts, and Hubert Wilhelm, all of whom made constructive suggestions which improved the volume. Hilda Kendron and Dorothy Tudanca supervised the typing, proofreading, and assembly of the manuscript. Several student assistants at the University of Akron labored at the usually thankless task of wordprocessing the manuscript. They include Christa J. Anderson, Julie A. Dohner, Lauren D. Downs, and Audra K. Wixom. The authors' special thanks go to Margaret Cleek, who learned to recognize German barns at fifty-five miles per hour, and Jane Noble, who happily resigned herself to never traveling anywhere by the most direct route.

The authors are grateful for permission to reprint the following photographs and drawings:

Figures 2.2, 3.2, 3.14, 3.23, 3.34, 4.10, 6.2, 6.4, 6.33, 6.44, 6.45, 7.4, 8.1, 8.5, 10.2, 10.3, 10.5, 11.2, 12.7, and 13.5 are reprinted from volume 2 of Allen G. Noble, *Wood, Brick, and Stone: The North American Settlement Landscape*, with drawings by M. Margaret Geib (Amherst: University of Massachusetts Press, 1984). Copyright © 1984 by The University of Massachusetts Press.

FIGURES 2.4, 6.35, 6.36, and 6.37 are reprinted from M. Comeaux, "The Cajun Barn," *Geographical Review*, vol. 79, no. 1 (January 1989): 49, 58. By permission of M. Comeaux and the American Geographical Society.

FIGURES 3.28, 3.30, 3.36, and 6.7 are reprinted from *Pioneer America* (March 1981): 29 (modified) and (July 1974): 13.

FIGURE 6.31 is reprinted from Hubert G. H. Wilhelm, "Amish-Mennonite Barns in Madison County, Ohio: The Persistence of Traditional Form Elements," *Ohio Geographer* 4 (1976): 5. By permission of the *Ohio Geographer: Recent Research Themes*.

FIGURE 6.34 is reprinted from Victor A. Konrad, "Against the Tide: French Canadian Barn Building Traditions in the St. John Valley of Maine," *American Review of Canadian Studies*, vol. 12, no. 2 (Summer 1982): 24.

FIGURE 6.46, by M. Geib, is reprinted from Allen G. Noble, ed., *To Build in a New Land* (Baltimore: Johns Hopkins University Press, 1992).

FIGURE 6.48 is reprinted from David R. Lee and Hector H. Lee, "Thatched Cowsheds of the Mormon Country," *Western Folklore* 40 (1981): 173. By permission of the California Folklore Society.

FIGURES 8.3 and 8.4 are by Jon T. Kilpinen, "The Mountain Horse Barn: A Case of Western Innovation," *P.A.S.T.: Pioneer America Society Transactions*, vol. 17 (1994): 26, 27.

FIGURE 9.1 is reprinted from John Passarello, "Adaptation of House Type to Changing Functions: A Sequence of Chicken House Styles in Petaluma," *California Geographer* 5 (1964): 70. By permission of the *California Geographer*.

The Old Barn Book

1

Introduction

This book is the most comprehensive ever written on North American barns and farm structures. Even here, however, not all types of buildings have been included. Often we excluded structures because they were nondescript or because their functions were not clearly enough defined. A case in point is the sheep fold. While designs for barns specifically intended to house sheep exist from about the turn of the century, these barns were never constructed in large numbers, and, to the present day, sheep are often housed in a variety of barns and miscellaneous nondescript buildings.

Furthermore, we have not studied every type of farm structure adequately, nor have we even identified each as distinctive. Those that have been examined in detail often have variants which have not yet been recognized and identified, especially if they are located in areas far from the original source of the building.

WHAT, WHERE, AND WHY

This book examines barns and farm structures in most parts of North America. Some areas such as northern Canada, the Adirondack Mountains of New York State, and the desert Southwest, to mention just three areas out of a much larger number, never had the agricultural development necessary to support many farm buildings. Elsewhere, an often bewildering array of structures awaits investigation. Even in the urbanized fringes of the great metropolitan centers, a few agricultural survivors often can be found, although for most of them their days are clearly numbered.

Both authors of this volume have traveled extensively across Canada and the United States in search of rural structures. Nevertheless, only a small portion of the two countries has been covered. We have attempted to expand the expertise of this volume by incorporating information from hundreds of studies by other investigators. Many are listed in the bibliography at the end of this book (see "Sources for More Information").

The rural structures of the eastern parts of Canada and the United States have received the most careful study to date. These areas experienced early settlement by the largest number of European immigrants, so structure variety stemming from ethnic diversity is greatest in the East. Settlement of central and western North America occurred later, after barns became standardized, with the hewn timber of earlier days being replaced by inexpensive sawn lumber. Designs and building techniques, such as the widespread use of standardized lumber trusses for roof support, also reduced building variety. Although far fewer studies have been made of barns and other farm structures in central and western North America than in the eastern part of the continent, in the future additional information will undoubtedly become available for these areas.

This book answers the most commonly encountered questions about rural structures and explains why features occur where they do. But not every question *can* be answered. For example, most barns have vertical siding. The horizontal siding on barns in the Shenandoah valley of Virginia is a result of the rebuilding of these structures by house carpenters after the destruction of the Civil War. Why horizontal siding is used on barns in the Mohawk valley of New York and elsewhere has never received an adequate explanation. Further study may at some time provide answers, but definite answers may never be supplied for some questions. Speculation plays a large part in the study of folk and vernacular buildings, and it is a large part of the fun of such study.

NORTH AMERICAN BARNS

Barns are among the most conspicuous and easily recognizable features of the countryside. Barns are visible links to a way of life that is rapidly disappearing. Not long ago, most

North Americans lived on farms or in small towns close to the countryside. Their acquaintance with barns was intimate—they knew them as workplaces, dance halls, social centers for husking bees and similar activities, and even as religious sites, much as the Amish continue to use them even today.

Modern-day Americans and Canadians know little of barns. Few have been inside a barn. Even fewer have experienced the pleasant smell of curing hay and the other more vivid, if less attractive, smells associated with the barn.

Because many barns are so large and often stand in relative isolation, they give particular character to the countryside. Barns in New England are different from those of the Midwest, which in turn are quite distinct from those of Appalachia, and so on. Even within regions, the form of the structures varies from place to place. Furthermore, different types of barns were introduced by different ethnic groups. Thus they identify the people whose descendants often still occupy the same farms.

The form of the barn also varies according to the agricultural function that the building was designed to perform. Animal barns are different from crop barns. A hay barn has a very different form from a potato barn, and each can usually be recognized and correctly identified if one knows what to look for. This guide provides the information to identify barns and the other common farm structures.

Regardless of the type, function, or location of the barn, certain common features occur in virtually all these structures. Almost all barns are rectangular, although a few built in the latter half of the nineteenth century, and the first two or three decades of the twentieth century, were circular or geometric in floor plan. Most of these are in the Midwest, but occasionally one encounters such a barn in other sections of the country, where often it is a local landmark and a subject of much local pride.

Barns dating from before the end of the eighteenth century (1700s) are very rare. The few that survive are not at all typical, because they usually are large, well built, and associated with the wealthiest properties. Still, we can visualize the typical early barn, given the many nineteenth-century structures that do survive, because most barn designs did not change much from the seventeenth up to the early twentieth century.

Where soils are not very fertile and the land hilly and covered with forest, farming never was very prosperous. In these

Fig. 1.1 A log, Flue-Cured Tobacco barn in southern Virginia.

areas, log was the preferred building material into the twentieth century.

The earliest center of log building in the eastern United States was south-central Pennsylvania. From here, such building spread across a large area where conditions were suitable. Later, a smaller area in the upper Great Lakes developed with the migration directly from Europe of other peoples who also possessed log-building skills.

Elsewhere, timber frame barns were built, again because of the ready availability of timber from vast forests. The timber frame barn used large posts and beams as a framework onto which siding and a roof were placed.

The choice between log and timber frame also depended to an extent on the building skills and traditions of those who built the barns. Germans and other central and northern Europeans were used to erecting log structures. The English, on the other hand, had for a long time nurtured a tradition of timber frame building. Those intrepid frontiersmen the Scotch-Irish did not originally build log buildings, but quickly picked up the technique from contact with Finns, Swedes, and Germans.

Barns built of brick and stone are much less common than

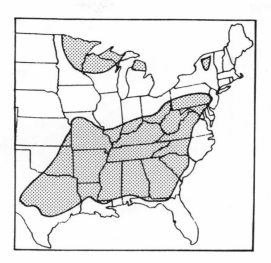

Fig. 1.2 General area of eastern United States where log barns may be encountered frequently.

Fig. 1.3 A stone barn with wooden frame doors from the Oley valley of southeastern Pennsylvania.

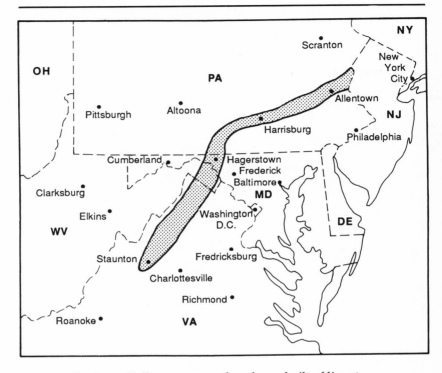

Fig. 1.4 The Great Valley, one area where barns built of limestone are found.

wooden barns, for several reasons. First, these materials are more expensive because considerable skill is required to prepare them: bricks must be shaped and fired; stone must be cut and dressed. Additional skill is needed to lay up walls with them. Second, because both brick and stone are heavy, they are rarely employed outside their areas of origin.

A geological map provides an excellent tool for searching out areas likely to have stone and brick barns. Limestone is the most common cut stone used in barns; sandstone is more difficult to cut and work with, and it has a more limited distribution.

Perhaps the best-known area of limestone barns in North America extends along the structural depression called the Great Valley, which runs from the Shenandoah valley of northern Virginia to the Lehigh valley of Pennsylvania.

Brick barns appear most often in areas of shale rock, which

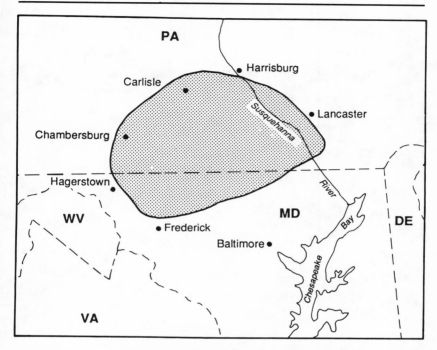

Fig. 1.5 This area has the highest density of Decorated Brick-End barns in the United States.

weathers to a clay soil from which bricks can be made. Probably the best-known area of brick barns occurs in south-central Pennsylvania, extending into northern Maryland. Many of these barns have decorated gable ends, created by leaving a few bricks out to form a geometric pattern.

The geology of an area can be roughly determined by looking at the foundations of barns, even timber frame ones. Limestone is often used, and it can be gray, bluish, or whitish; sandstone is usually tan or yellowish in color. Shale is not often used for foundations but bricks may be, which would indicate the presence of shale rock. In glaciated areas, granite boulders and rounded metamorphic rocks are laid up in a mortar filling.

Barn (and house) foundations often indicate where the geology of an area changes, where one crosses from one type of rock to another. Because stone and bricks are heavy, they are used primarily near their source.

Fig. 1.6 A typical Decorated Brick-End barn from near Carlisle, Pennsylvania. The diamond-shaped patterns are created by leaving out certain bricks to create ventilation. The technique is probably a melding of British and German traditions.

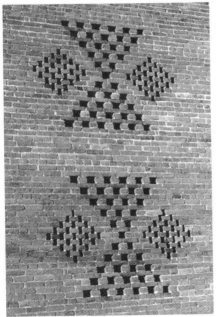

Fig. 1.7 Closeup of brick patterns from a Decorated Brick-End barn in York County, Pennsylvania.

BARN LOCATION

What makes the search to identify farm structures exciting is that their location is only partly predictable. Many early farmers built their barns from plans and ideas stored in their heads. The farming culture was so much a part of them that they instinctively knew what a barn should look like and how it should be built. Nearly everybody in their community built barns the same way, and furthermore, they had helped others build barns. The well-known Amish barn raisings are a hold-over of this traditional approach.

Gradually some especially talented carpenters began to specialize in building barns for others, but they still conformed to the norms of the community. German builders built German barns, often easily recognized because of their forebay overhang. Dutch builders erected gable-entry barns with huge interior anchor beams which immediately marked them as Dutch. Other builders produced barns with their own distinctive ethnic trademark.

As the rural countryside filled with people in the nineteenth century, the different ethnic groups came into more intimate contact with one another and they began to exchange ideas. The lines of cultural differentiation became less clearly defined, especially as agriculture became more scientific and required buildings that were more efficient. The result was that barn buildings slowly became more and more alike.

Ideas of how to build a barn have always traveled with people, so barns today tell us about yesterday's settlement. We can see this development clearly in the Tuscarawas and other valleys of eastern Ohio where German Bank barns and English Raised barns define the fertile soils of the river valley which early attracted the Germans, English, and Scotch-Irish. Small English threshing barns and Double-Crib barns dot the uplands, where only subsistence farming is practiced. These lands had to content the later-arriving settlers. The presence of the two barn types in Tuscarawas County identifies not only the major rural ethnic communities of the county, but also the quality of the farmland.

Another example can be drawn from southeastern Appalachia, where a series of disparate topographic provinces lies in parallel northeast-to-southwest bands. Because the physical characteristics, and consequently the agricultural potential, within each of these areas vary so greatly, sharp contrasts

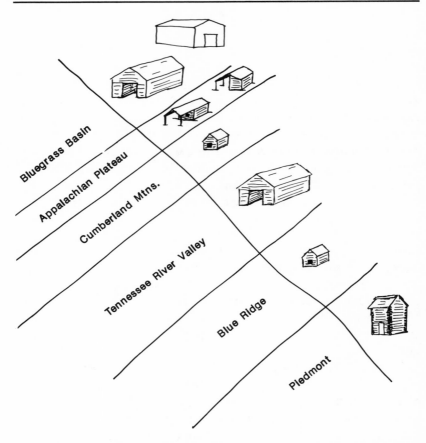

Fig. 1.8 Variation in types of barns from the Bluegrass Basin of Kentucky to the Piedmont of North Carolina. Areas are not drawn to scale.

exist in the types of barns encountered in each area. In the Bluegrass Basin of central Kentucky, Transverse Frame barns predominate. Many of these have been altered by the addition of narrow wall ventilators and long, ridge-top ventilators to aid tobacco drying. The barns are invariably painted with a creosote protective covering to an almost black color.

Moving southeastward into the rolling hills of the Appalachian Plateau, the number and size of Transverse Frame barns decline, Tobacco barns disappear, and Side-Drive and Front-Drive Crib barns appear. Farther southeast in the Cumberland Mountains, where the topography is rugged and little agriculture is carried on, only simple (i.e., primitive) crib barns occur.

The wide valley of the Tennessee River offers the opposite conditions—gentle slope and fertile limestone-based soils. Consequently, the Transverse Frame barn dominates, although a scattering of various crib barns also can be seen. Next, to the southeast, the Blue Ridge Mountains bring a return of poor agriculture and crude crib barns. Finally, on the Piedmont of North Carolina, the tall, squarish unpainted forms of the Flue-Cured Tobacco barns dot the rolling countryside. Clearly, in this part of the United States, the distribution of barn types is closely related to topography and soil productivity. The same can be said of many other areas.

Another example from southern Wisconsin demonstrates the effects of ethnicity and agricultural change. English barns are found within a mile or two of small communities, relicts of early British and Yankee settlers and an early economy based on wheat. Farther from town are the larger Raised barns, preferred by the later-arriving Germans; these barns were more efficient as the economy turned to dairying.

A final example from southwestern Iowa illustrates another variation on the basic theme. Along U.S. Highway 36 stands a huge, majestic German Bank barn surrounded by dozens of farms on which the barns are of the transverse frame, Saxon type. The Bank barn must have its origins in the westward migration of a farm family whose North American roots lay in Pennsylvania (see fig 6.12 for a definition of this pathway of migration) and earlier, its European roots in southern Germany. Most of the farmers of southwestern Iowa, however, trace their heritage more directly to Germany, and especially to the North German Plain. The lone south German barn stands out in a sea of north German barns and gives us a rich lesson on ethnic origins and migration.

THE SOCIAL HISTORY OF BARNS

The barn is a working building. It can house animals, store harvests, and provide a work area for accomplishing various farm tasks, although not all barns provide all of these functions. Given the utility of a barn, it is perhaps surprising that barns are uncommon outside of Europe and the European-settled areas of the world.

The lack of barns in much of the world can be explained, in part, by warm climates in which livestock do not have to be sheltered; climate may account for the paucity of barns in southern Europe and the southern United States. Also, many cultures do not keep large animals. Even those that do may graze them over wide areas without sheltering buildings, as did turn-of-the-century cattle ranchers in western North America. Finally, a disturbingly large number of the earth's rural inhabitants do not produce the agricultural surplus that would necessitate the large-scale storage and processing for which the barn is designed.

In Britain the Romans built barns whose traces are still discernable. Later, during the thirteenth century, large ecclesiastic tithe barns were constructed in Europe; some still exist today. Most European barns that have been studied, however, date from the sixteenth century or later. Certainly this dearth of earlier barns may simply reflect the fact that very old barns vanish through decay and destruction. Still it is probable that the building of large numbers of substantial barns may date only from the sixteenth century, the beginnings of the second agricultural revolution.

This period in European history is marked by expanding populations, in part a resurgence following the ravages of the Black Death, and in part a reflection of the transfer of communal landholdings to private ownership. Later, beginning in the seventeenth century, improved methods of crop rotation, fertilization, and innovations in agricultural tools and machinery denoted a transition from subsistence to commercial agriculture. Higher crop yields and larger herds necessitated the widespread construction of buildings to provide for crop storage and animal shelter. The increased agricultural prosperity in England also contributed to the construction that has been termed the Great Rebuilding.

In America the adage "It doesn't do any good to lock the barn after the cows are out" replaced the older British version, "Lock the stable before the horse is stolen." This difference reflects more than linguistic variation. A typical, although not universal, farm layout in Britain includes a barn for crop storage and processing, a "cow house" for housing cattle, and a stable for horses. Americans typically combine crop storage, processing, cattle housing, and even horse stabling into the barn.

The typical British farm pattern described above obviously did not diffuse widely in North America, which raises several logical questions. Why not? Which barns did settlers bring with them from the Old World? Which barns did they invent to meet the environmental, economic, and social conditions of their new homeland? What melding of old and new occurred as the English, Scotch, Irish, German, Scandinavian, and Dutch settled deciduous or coniferous forests, rocky and glaciated uplands, fertile river valleys, and grassy plains? What patterns were introduced by the later waves of Bohemian, Ukrainian, Belgian, and Polish farmers? Did farmers from Massachusetts or Tennessee or Ontario take their barn types with them when they moved to California or Wisconsin or Saskatchewan?

Some of these questions are addressed by cultural geographers as well as folklorist and agricultural historians, while others remain to be answered. We know that the European colonists, with some exceptions, did not bring with them their large housebarns, so common throughout the old countries. Pioneer settlers, faced with clearing virgin forest or breaking sod, had little time to do more than erect a rough house and perhaps a crude animal shelter in the early years. Often, unlike the patterns of the old country, the family was small, nuclear, and living on its own land. An ox or two, perhaps a horse, and one or two cows constituted the larger livestock, with the cows often allowed to free-range. Not until some ten years after settlement, or perhaps not even until the second generation, did the pioneer have the wherewithal and the need for a barn. This description may not apply to the Germans, who are mentioned regularly in early reports as building their barns better than their houses and earlier than their neighbors. The paucity of barns and their rough quality in the Midwest and West were especially noticeable.

Log construction of both houses and barns predominated on the frontier. The material was abundant, and the techniques of construction were known by the Scandinavians, Finns, Swiss, and south Germans and borrowed by the other ethnic groups who had little background in log construction. While some Old World log-barn types were re-created in the New World, many of the North American log-barn structures were simple, boxlike, utilitarian cribs. Often within a generation or two, log construction was replaced with timber framing and stone or brick, allowing barns like those of the homeland to be built. Clear European antecedents exist for a number of American barn types, and English, German, Swiss, Dutch, and French colonists transplanted some of their barns onto New World soil with only minor modification.

Ethnicity explains only a part of the inventory of North American barns. As utilitarian buildings, barns reflect the crops and animals of a time and place. The needs of a pioneer wheat farmer for a small barn with a threshing floor are quite different from the needs of a dairy farmer with a herd that must be fed and housed during a long winter. Tobacco requires storage unlike that demanded for potatoes or hops. The demise of wheat in older, settled areas as new lands were opened, the innovation of corn-fattened beef and hogs, and the introduction of large-scale dairying all had impacts on the barn landscape.

Barns reflect technological changes. The advent of the hay baler, the thresher, and the silo clearly affected barn design, as did the invention of the overhead hay carrier and the hay fork. Steam power, the gasoline engine, and electricity also had radical implications for farm operations. By the early 1900s, barn plans and even whole barns could be ordered by catalog. In the twentieth century, timber framing was rarely used, replaced by plank framing. Gable roofs were increasingly supplanted by gambrel and Gothic roofs. Like most technology, agricultural technology continues to develop, sometimes resulting in major changes in the farm landscape. An aging former Kansas farmboy, now living in urban America, would be surprised at the large green circles of irrigated crops so visible from airplanes traveling over the Great Plains, or the huge round hay bales, the horizontal, tire-laden silos, or the rubber-strip fences which have recently become part of the agricultural landscape of the Midwest.

Time and technological changes have made the barn increasingly scarce in this landscape. Yet for most North Americans, the barn has become more than a vanishing, utilitarian building. It has become entrenched in our cultural symbolism. Our language still includes such expressions as "can't hit the broad side of a barn," "big as a barn door," and "barnyard humor." A sensational speech is a "barn burner," and politicians still "barnstorm" across the country. Even the most urban American realizes something of the bygone social and cultural implications of a "barn raising." Sadly, our grandchildren may know of barns only through these increasingly archaic figures of speech and through open-air agricultural museums.

TERRA INCOGNITA

Terra Incognita—the unknown lands of old maps were often so inscribed. The "barn map" of North America contains such lands as well. With a roughly agreed-upon barn typology just recently set out, and faced with a vast continent, approached by hardly more than a score of barn scholars, it is perhaps not surprising that most of North America has not yet been systematically inventoried. Indeed, some areas have not even been sampled. Compounding this situation is the increasingly rapid disappearance of many of the structures described in this book, or worse, the disappearance of undescribed types and structures, for surely they exist.

A solution lies in enlisting your aid. Remember, for example, that nonprofessional scholars, observers, and enthusiasts have made major contributions to our understanding of North America birds and, closer to our topic, to the preservation of historic architecture. With public participation we can achieve some of that same success in understanding barns and other rural structures and, in the process, understanding our own past.

If you see a barn or farm structure that is outside its "range" or a type not described in this guide, we urge you to submit a report to the North American Barn Project. While we cannot guarantee an answer to all correspondence, we can serve as a repository of information that can be of immense aid to future scholars.

At the back of the book are several reporting forms. Please fill them out as completely as possible, although we recognize that not all information will be available to you. Try especially to provide an accurate road location. Specific roof types, floor plans, and constructional features are described in different sections of this guide. Wagon-door placement is a key diagnostic feature. Please make clear whether the larger doors are on the side or end of the barn, and centered or off-center. Note the placement in a sketch as well. Is the foundation nonvisible, low, or full height? Also, is the foundation made of fieldstone, concrete, cut stone, or brick? What about walls?

Typical ventilation and decorative features also should be noted. Please illustrate these in your sketch.

Photographs are of great help. Please try to fill the frame of your viewfinder with the barn. Two shots, each showing a side and end, will capture most barns. Please sketch your barn and its key features on the back of the form. Sketches can be valuable, even if you submit photographs.

Finally, provide, if you can, the name and address of the building's current owner. Also give any information available on the history of the structure.

DATING OLD BARNS

Providing an accurate and precise time of construction for barns is usually difficult. The older the barn the more difficult. Written records often do not exist to confirm a construction date, memories of current family members are imprecise (one hundred years old is a frequently related age for any old building), and often the farm is no longer in the same family as the builder of the barn. Occasionally a date stone or a date inscribed in a post or beam or some other physical evidence is encountered, but not often.

A few clues help to estimate age. Low, Pole barns were almost always constructed post–World War II. Gothic or Round-Roof barns were built predominantly between the world wars. Barns constructed of sawn lumber frame were rarely built after 1870, but a few were built as late as the mid-twentieth century. Gable roof barns are usually older than barns with other types of roofs, but not always. The gambrel roof style was popular from just after the Civil War to World War II. A steep roof pitch often indicates an older barn, whose design dates from the time when thatch was a common roofing material.

Sometimes property deeds or local newspapers can provide clues to date of construction, but tracking the information down is tedious and time-consuming.

The Barn

2

An Overview
of Barn Types

Part 1 of this volume identifies the key features of barns and something about their locations. Then, arranged in roughly chronological sequence, are descriptions of more than fifty barn types, beginning with the *crib barns* and *crib-derived barns*, proceeding through the large array of *ethnic barns* to the *more recent types*, and ending with *special-purpose barns*.

Learning to identify the barn in front of you may seem a daunting task at first. But, by examining the critical features of the barn, you will be able to strip away the extraneous detail and focus on the diagnostic features—features often encumbered with additions and modifications, as over the years, farmers have adapted to changing conditions.

GEOGRAPHICAL PATTERNS

A barn's location can be a valuable clue to identification. Most barn types fall into regional patterns. Some types are so localized that they are found only within one or two counties of a single state. On the other hand, a few types are distributed across North America, with varying frequency. We have provided "range" entries for each barn type, listing areas where the barn is known to occur. You should consult part 3, "Where and What," at the end of the book, to learn what sort of structure to look for in a particular geographical area. A few types have been mapped, and we have provided those maps where possible.

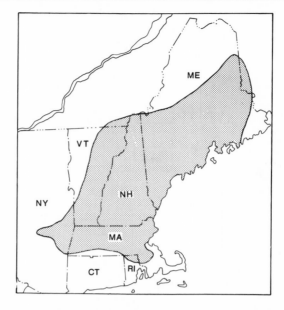

Fig. 2.1 Areas where New England Connected barns can be found.

BARN TYPES

CRIB BARNS

Crib barns, like log cabins, were probably ubiquitous where frontier and forest coincided because they were relatively easy to construct from logs already cut in the process of clearing land. But farmers always needed more space for animals or crop storage or both, so there are several elaborations of the basic Single-Crib barn. Today, crib barns are scarce. Often the earliest barns in an area, they were usually crudely built and have since deteriorated. As the frontier period ended and farms became larger and better established, ever larger structures with more storage space were needed, giving rise to the crib-derived barns. These barns are timber-framed rather than constructed as log cribs, but their floor plans clearly reflect their crib-barn origins.

ETHNIC BARNS

The majority of the barns in this book are ethnic barns. Some are closely linked to an ethnic group and even have anteced-

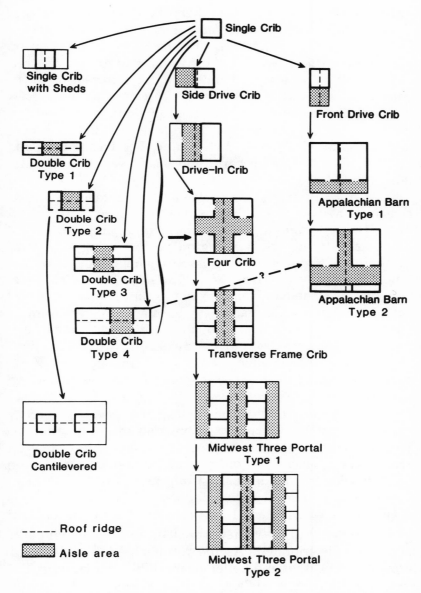

Single Crib

Single Crib
with Sheds

Side Drive Crib

Front Drive Crib

Double Crib
Type 1

Drive-In Crib

Appalachian Barn
Type 1

Double Crib
Type 2

Four Crib

?

Double Crib
Type 3

Appalachian Barn
Type 2

Double Crib
Type 4

Transverse Frame Crib

Double Crib
Cantilevered

Midwest Three Portal
Type 1

----- Roof ridge

▨ Aisle area

Midwest Three Portal
Type 2

Fig. 2.2 Conjectured evolution of crib barn floor plans. (Courtesy of M. Margaret Geib and the University of Massachusetts Press)

ents in the old country. For others the ethnic links are less definite, and much work remains to be done to firmly establish their ethnic connections. We begin with the English, among the earliest farmers of the immigrant groups and certainly the most widespread and enduring ethnic group of the early colonial period. The Three-Bay Threshing barn, the New England Connected barn, the English Bank barn, and the Welsh Gable-Entry barn are all found, albeit with some modification, on both sides of the Atlantic. The Raised barn and certainly the Foundation barn represent a New World evolution of these earlier English barns.

The German migration, a trickle in the eighteenth century, swelled to a flood during the nineteenth century. German ethnicity is predominant from Pennsylvania across the Great Lakes states to western Washington and extending southward through Colorado, Kansas, Missouri, and Illinois. Concentrated outliers exist in New York, Texas, northern California, Oregon, the Canadian Plains provinces, and Ontario. It is not surprising that the Germans have been associated with more than fifteen types of barns. However, only the Grundscheier (ground barn) varieties and the Sweitzer (Swiss) barn have had clear antecedents identified in the Germanic areas of Europe. The myriad other types that we associate with German settlement appear to have evolved in southeastern Pennsylvania.

Most of these German barns, at least ten types, are forebay barns and can be difficult to distinguish from one another. The most common forms are those in which the forebay is an extension of the second floor beyond the foundation wall on the downslope side of the barn. Confronted with such a barn, the observer should note three things: whether the roofline is symmetrical; whether and how the forebay is supported; and what is the depth of the forebay. These factors should allow the identification of this most elusive category of barns.

At least twenty major ethnic groups, and perhaps another twenty less populous ones, could have brought their barns to North America, that is, they were agriculturalists from temperate or cold climates. Yet, including the English and Germans noted earlier, we find only eleven groups that have been closely identified with one or more North American barn types. This probably reflects a lack of research and study as much as a failure of a particular immigrant group to transplant its barns or to create a new type in a new environment.

Fig. 2.3 *Grundscheier with Unbroken Forebay located in the Oley valley, Pennsylvania.*

Fig. 2.4 *A typical Cajun barn located near Arnaudville, Louisiana. The Cajuns are one ethnic group whose structures are recently beginning to be well documented. (Courtesy of M. Comeaux and the American Geographical Society)*

Fig. 2.5 A Raised Round barn located in northeastern Indiana.

The Dutch, French, Swedes, Finns, Czechs, and Hispanics built barns that often identify their respective ethnic landscapes. So did the Amish, German-Russian Mennonites, and Mormons. One wonders whether there are as yet undiscovered Norwegian, Danish, Belgian, Slovenian, Bohemian, or Russian barns. If not, why did these groups adopt existing barn types and abandon their own?

MORE RECENT BARNS

Most recent barns date from the last decades of the nineteenth century. Their creation reflects a variety of factors. More space for increasing herd size and agricultural processing was provided by the Three-End barn. Round and Polygonal barns were promoted as more efficient designs by some popular lecturers and writers, as were octagonal houses. The Erie Shore barn provided more light and ventilation for the small farmer.

All of these types took advantage of new building technologies, such as balloon framing with dimension lumber, and new machinery, such as the hay track.

SPECIAL-PURPOSE BARNS

Special-purpose barns were usually designed and built for a particular agricultural operation or situation and therefore are quite distinctive. The Wisconsin Dairy barn, with ample light and ventilation, incorporates the scientific knowledge of the turn of the century and was actively promoted by the University of Wisconsin's school of agriculture. The Mountain Horse barn reflects both the available building material and need for animal shelter in high elevations of the Rocky Mountains.

Certain barns evolved distinctive forms became they were employed to store or process a single farm product. Half-sunken Potato barns meet the special storage requirements of this root crop. Hop barns provide an efficient solution to drying and storing of the hop pods. Each of the three methods of drying tobacco leaves is associated with a major type of Tobacco barn, although much variation exists within each type.

IDENTIFYING BARNS

The careful observer can find his or her way through these barn types and identify most barns by asking some key questions.

How many levels or floors does the barn have?
Does the barn have second-floor access via a bank or ramp?
Is the floor plan square or rectangular, or some other shape?
Is there a forebay?
Is there a large wagon door, and where is it placed?

The answers to these questions should result in a small set of possible barns. Closer examination should produce a final identification.

3

Barn Features

PARTS OF A BARN

Although fast disappearing, the most common type of barn construction in eastern North America is timber frame, so-called because it consists of a heavy frame of hewn timbers. In England such construction is usually referred to as box frame, because without considering the roof, the timbers form a rectangular box. In North America this type of construction is also called *post and beam,* in which all vertical structural members are kinds of *posts* and all horizontal members are types of *beams,* although some may have more particular names. The four beams at the base of the walls next to the foundation are called *sills.* The beams at the tops of the side walls, underneath the *eaves,* are called *plates.* Every barn has at least four sills and two plates. Other beams in the plane of the walls are called *girts.* Beams connecting the front and the back sides of the barn, but not always in the plane of the walls, are called *tie beams.* Posts are identified by the various positions they occupy in the frame, namely, *corner posts, end posts, side posts,* and *interior posts.* Short diagonals that connect posts and beams are called *braces.*

Posts, beams, and braces are held together by wooden pegs pounded into holes bored through mortise and tenon joints. These pegged joints are often much stronger than joints held together by nails or spikes. Some pegged joints are quite complicated, truly carpenters' works of art, although joints in

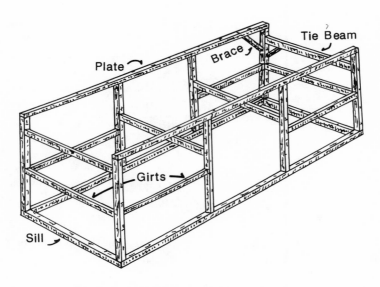

Fig. 3.1 The basic structure of a timber frame barn.

North American barns are not nearly as elaborate as some of those found in European barns.

The frame of the barn is usually put together on the ground in sections called *bents*. Once assembled, they are raised into place and connected by girts to other bents. In Dutch barns girts are called *struts*. The distance between bents is called a *bay*. Three-bay-wide and two-bay-deep barns are probably the most common of older North American barns.

The normal front and back of the barn, that is to say, the walls under the *eaves* of the barn, are more properly referred to as the *sides*. The other walls are the *gable ends*, or simply *ends*. The *gable* properly is the upper triangular area from the *ridge* to the eaves.

Two major divisions of North American barns can be identified according to whether the doors are on the side wall or the gable end of the structure. As a general rule, barns in northeastern United States and adjacent Canada have main doors on the barn side, while in southeastern United States main doors are on the gable. In western Canada and the United States no clearcut division exists.

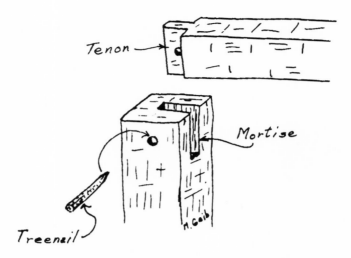

Fig. 3.2 Detail of a mortise-and-tenon joint used in the construction of timber frame barns. (Courtesy of M. Margaret Geib and the University of Massachusetts Press)

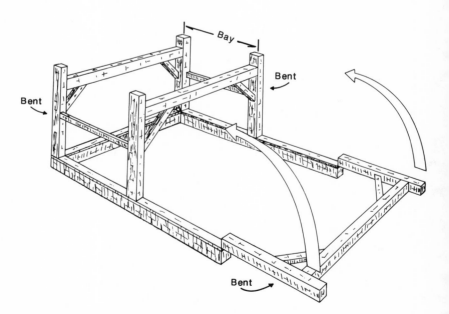

Fig. 3.3 Bents and bays of a timber frame barn, suggesting the method of erecting the frame of many barns.

Fig. 3.4 *The exterior parts of a timber frame barn.*

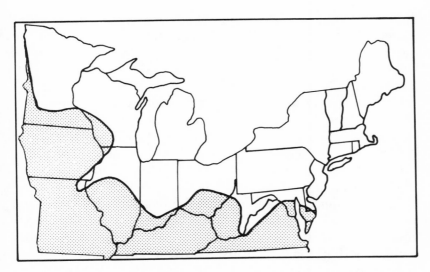

Fig. 3.5 *Generalized boundary between barns with doors on the side and barns with doors on the gable. The shaded area represents barns with gable doors.*

Door on Side	Door on End
Double-Crib	Single-Crib
Appalachian	Drive-in Crib
English	Four-Crib
Three-Bay Threshing	Transverse Frame
Raised	Midwest Three-Portal
German	Feeder
Swedish	Welsh Gable-Entry Banked
Quebec Long	Welsh One-Story
Acadian	Dutch
Three-End	Finnish Cattle
Erie Shore	Finnish Meadow Hay
Gothic or Round-Roof	Madawaska Twin
Pole	Cajun
Potato	Czech
Tobacco barns (most)	Wisconsin Dairy

WALL MATERIALS AND FOUNDATIONS

WALL MATERIALS

A few barns are made entirely of stone or brick. Many older barns may be cribs of horizontal log construction. Often the logs are sided with planks and are no longer visible. Most barns, however, are of timber frame, sided with planks applied vertically. Vertical boards do not encourage rainwater to collect on top surfaces and thereby guard against wood rot. Horizontal planking is less common, but occurs in New England, among Dutch barns of New York and New Jersey, and occasionally elsewhere. Wisconsin counties settled after the Civil War have large numbers of horizontally sided barns. In the Shenandoah valley a significant number of barns were rebuilt on earlier foundations after the destruction of the Civil War. The work was done by house carpenters who used horizontal rather than vertical cladding because they were more familiar with it. Walls comprised of shakes of cedar or pine are encountered in coastal New England (especially Maine) and in New Jersey (especially Monmouth and Somerset counties).

Variant 1. Half-timber walls (Fachwerk) are rare and found only in early barns of German origin. In most instances, later clapboards cover the half-timber construction, so it is difficult

Fig. 3.6 *Vertical cladding, the most common barn siding, on a crib barn from central Indiana. No attempt is made to make the walls completely weather-tight.*

to detect. A very steep roof pitch is another clue to look for if you suspect half-timbering.

Variant 2. The walls of some early lumber frame barns are filled with packed mud.
Range: Missouri and probably other parts of the Midwest.

Variant 3. Stovewood (stackwood, cordwood) walls are comprised of short lengths of wood piled up so that one end of each log is exposed.
Range: They are rare and have been reported from only a few locations near the upper Great Lakes and more widely in Canada.

Variant 4. In central and western New York, cobblestones are sometimes used as walling material. An even larger number of timber frame structures employ cobblestone foundations.
Range: Central and western New York, especially around Rochester; near Paris, Ontario, and around Beloit, Wisconsin.

FOUNDATIONS

Foundations usually consist of materials different from those that comprise the walls. Almost all farm buildings have some

Fig. 3.7 Example of Fachwerk in which the framework is timber and the in-filling is stuccoed mud. This early German stable is now at Old World Wisconsin Outdoor Museum in Eagle, Wisconsin.

Fig. 3.8 Barn wall composed of stovewood on a barn in northern Wisconsin. The 18-inch-long lengths of wood have been split into rough halves and laid up in a cement mortar.

Fig. 3.9 An example of a cobblestone foundation from Cortland County, New York. Courses of both herringbone and round cobbles are randomly employed.

sort of foundation. Even log cribs are normally set upon large corner fieldstones to raise the sills above the sometimes damp ground, so as to reduce rot. Barns of stone and brick may have foundations fully integrated with the walls.

Almost all stone barns fall into one of three categories, based on their material of construction: sandstone, limestone, or fieldstone. Fieldstone consists of rocks and boulders picked up from fields and is easily available in much of the northern United States and eastern Canada. Fieldstone barns are relatively uncommon; instead the material is more often used for foundations for Raised and Bank barns. Because fieldstone is most frequently used where it is most abundant—in glaciated areas—fieldstone structures are a good way to identify those areas. A few barns, mostly in New York State, have cobblestone walls or foundations.

Cut-stone and brick foundation materials are relatively uncommon. Concrete block foundation walls are common on more recent barns. Some apparent cut-stone foundation walls

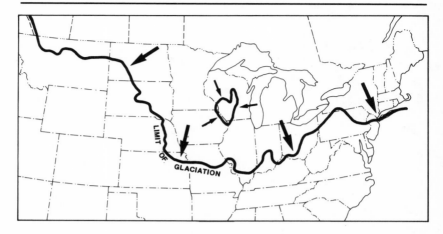

Fig. 3.10 Glaciated areas of central and eastern North America. Fieldstone is used for barn walls, and especially foundations, throughout this area.

in barns, silos, and houses are in fact concrete blocks molded to look like stone. Popular shortly after the turn of the century, these blocks were made in machines sold widely by Sears Roebuck. All-brick barns are unusual except in south-central Pennsylvania.

Fig. 3.11 Fieldstone foundation for a timber frame barn. Note the cut-stone corner quoins. Note also the vertical siding, which is the most typical wall covering for barns. This barn is in central New York.

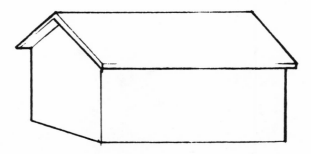

Fig. 3.12 The gable roof, also called a saddle roof or a ridge roof, is the most common type of barn roof.

ROOF TYPES

Roofs can help distinguish between barn types, but roof form by itself is not usually the determining factor in identification. In fact, when a barn is expanded or remodeled the type of roof may sometimes be changed, as, for example, when a gable roof is replaced by a gambrel. The major roof types are listed here.

GABLE ROOF

The earliest and simplest roof form is the gable. Very early gable roofs have a steep pitch (over 45 degrees) necessitated by the use of thatching when the roof design was conceived. Thatching did not long remain popular with migrants to North America, however, due to the long, cold winters, strong winds, and abundant wood for planks and shingles. Still, the steep pitch was retained for many years even though the roof material had changed to shingles, shakes, or planks.

In some types of barns, notably some of the Pennsylvania German barns, roof pitch actually increased with transfer of the type to the New World because thatch had not been used in Europe, but was used in early Pennsylvania. Noting door placement in the gable side or gable end can aid in more precise identification of barns with gable roofs.

Other terms: Saddle roof, ridge roof.

Range: Entire continent.

Variant. A "broken roof" appearance can result from sheds with different roof pitches attached to the sides of a barn.

Fig. 3.13 A broken-roof variant in which the sides have a gentler slope than the center. Note the difference from the gambrel roof (fig. 3.15), in which the opposite is true. This barn is near Maysville, in northern Kentucky.

Sometimes barns are constructed initially with this appearance (see Midwest Three-Portal barn). The key in recognizing this barn is noticing a change in pitch between the lower and the upper roof slopes.

Other term: Called a shed-roof barn in the central Midwest.

Range: Central Midwest especially, but also generally throughout western Canada and United States.

GAMBREL ROOF

This dual-pitch roof covers the same wall dimensions as a gable roof, but allows for much more loft storage space. Gambrel roofs often have been used to replace worn-out gable roofs. The gambrel is a more complex form than the gable and is more expensive to construct. A gambrel roof with a slight flare at the eaves is sometimes called a Dutch gambrel. The flare or "kick" throws rainwater away from the base of the wall, thus helping protect and preserve the barn's foundation. This roof became popular after the Civil War.

Range: General throughout continent, especially in eastern Midwest and southern Ontario.

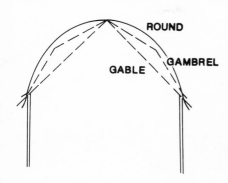

Fig. 3.14 Comparative loft area of gable, gambrel, and round roof barns. Increased loft area for the same floor space was one of the important reasons for the shift to gambrel roofs at the end of the nineteenth century and toward round roofs in the early twentieth century. (Courtesy of M. Margaret Geib and the University of Massachusetts Press)

ROUND ROOF

The popularity of the round roof dates from the 1920s. It is even more efficient than the gambrel in providing loft storage space (see fig. 3.14). The curve of this roof, often parabolic, can vary widely. Barns with this type of roof are not common, but they are widely distributed. The roofline is often broken by a tall entrance door.

Other terms: Gothic, arched, or rainbow roof.

Range: Widely and thinly scattered, most likely to be encountered in the Midwest, especially in areas of late settlement such as northern Wisconsin.

HIP ROOF OR HIPPED ROOF

The hip or hipped roof is also uncommon as a barn roof. More often it is seen on smaller outbuildings, especially in the

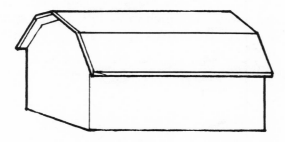

Fig. 3.15 The gambrel roof. This roof receives its name from the resemblance of the roof profile to a butcher's hook, called a gambrel *in French.*

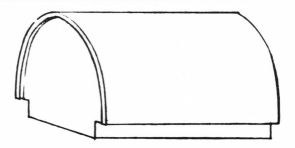

Fig. 3.16 A Gothic, arched, or round roof—sometimes also called a rainbow roof.

South. The hip roof has a roof slope on all four sides. It is the least expensive roof to construct, but offers the least loft area. Range: Widespread, but sparse.

Variants. Hip on gable, in which the upper part of the roof is gable and the lower is hipped; snug Dutch, in which the roof is gable with its upper corners cut off. Pyramid roofs, in which there is no ridgeline, occur on small buildings.

SALTBOX ROOF
Gable-roofed barns with one attached side shed often give the appearance of having a saltbox roof, if the shed roof continues the pitch of the barn roof. Some barns were built originally with this roof form (see Southwest Michigan Dutch barn). Range: One common area is southwestern Michigan. They are also scattered elsewhere.

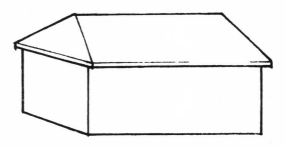

Fig. 3.17 The hip roof has the disadvantage of reduced loft space, so it is not very popular as a barn roof.

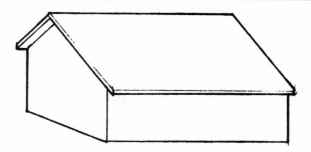

Fig. 3.18 The saltbox roof has one slope much longer than the other, although generally both slopes have the same pitch. A saltbox profile often results from a later addition to one side of a barn.

MONITOR ROOF

The monitor roof provides loft storage that is intermediate between a gable and gambrel roof, and is more common on smaller outbuildings than barns. A barn with flanking sheds may present this roof appearance if the sheds are shorter than the barn. It is sometimes called a clerestory roof.

Range: Common throughout California interior areas. It is also found in the Midwest, Appalachia and southeastern United States, Louisiana Cajun areas, and the Winooski River basin in Vermont.

PENT ROOF

The pent roof is not a barn roof proper but an unsupported roof extension attached to a barn wall to provide additional

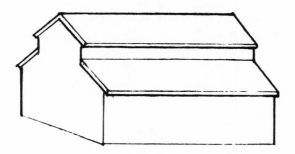

Fig. 3.19 The monitor roof, in which the center section is raised to provide better light and sometimes better ventilation. It is also called a clerestory roof.

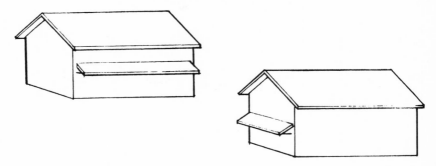

Fig. 3.20 The pent roof may be attached to either the barn end or to the barn side.

outside shelter. Pent roofs attached to the side of a barn may have some connection to the cantilevered overhang of the German Bank barn. Pent roofs attached to the gable end of a barn, however, probably come from an English tradition. The addition of sheds to the gable ends of English barns, which is common in the eastern Midwest, may be an outgrowth of the gable pent roof since this latter feature is common there.

Range: Gable-end pents are especially common in southern Michigan, eastern Indiana, and western Ohio on one-and-a-half-story barns and in north-central Ohio on Raised or two-and-a-half story barns. Also found in lower Mohawk and Schoharie valleys in New York.

PENTICE

The pentice is a variation of the full pent roof, but smaller and limited to the area over the main barn doors. Common on Dutch barns, it is also found on some English and German barns.

Range: Middle Atlantic states, especially New York. Less common in eastern Midwest.

HAY HOODS

Hay hoods are extensions at the ridge of the barn roof which protect or support pulley attachments used to load hay into the loft. Sometimes they also provide weather protection for

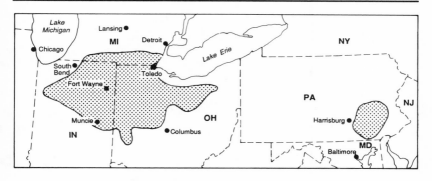

Fig. 3.21 Areas in which gable-end pent roofs are commonly found. Isolated examples occur over a wider area. The distribution of side pents is more difficult to identify.

the loft door. They occur in a variety of shapes ranging from a simple pole extension to complex box structures. Simple triangular hay hoods are often called hanging gables. Most common on crib-derived barns and on the more recent barn

Fig. 3.22 Example of a pentice on a Dutch barn in Montgomery County, New York. Note also the horizontal barn siding which is typical of most Dutch barns.

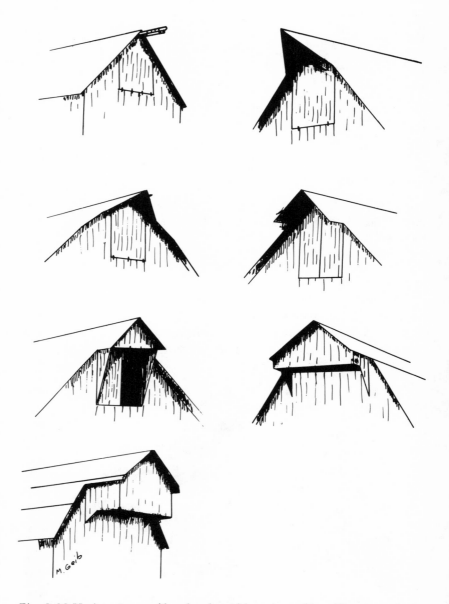

Fig. 3.23 Various types of hay hoods and hanging gables. (Courtesy of M. Margaret Geib and the University of Massachusetts Press)

designs, all of which tend to have gable-end loft openings, hay hoods offer one of the most easily identified barn features.

Other terms: Hanging gable, hay bonnet.

VENTILATORS

GABLE VENTILATOR

Lofts require ventilation, especially if used to store hay, which is subject to spontaneous combustion. In Appalachia the uppermost triangle of the gable wall is often left unboarded, especially on the end facing away from the prevailing wind.

Range: Widespread in Appalachia, especially central and southern parts.

Variant. The Lancaster ventilator is an overlapping triangle of siding at the line of the barge board which extends 6–8 inches beyond the line of the wall. It protects the gable-peak opening from rain and snow but still allows air to circulate.

Fig. 3.24 Lancaster-type gable ventilator has an open upper wall protected by an extended bargeboard. This barn is located in Champaign County, Ohio.

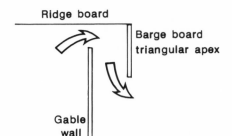

Fig. 3.25 Diagram showing air flow in the Lancaster ventilator.

Range: Lancaster County, Pennsylvania, west-central Ohio, and perhaps elsewhere.

CUPOLA RIDGE VENTILATOR
Architecturally impressive cupolas are a feature of many German Bank barns and Raised barns. Often the four sides of square cupolas are pierced by Victorian-style louvers, and the cupolas are capped with fancy roofs and lightning rods.
Range: Dairy belt of northern Midwest, northeastern United States, and eastern Canada.

Variant. In the twentieth century before World War II, steel ventilators were added to many dairy barns. Many of these ventilators contained metal fans activated by rising hot air generated within the loft. The idea did not catch on, and later barns do not have this feature.

TOBACCO BARN VENTILATORS
Tobacco barn ventilators may have a raised ridge, a series of metal ridge vents, or wall panels which can be opened. Further details on Tobacco barns are given in chapter 8.

DORMERS

Dormers interrupt the roofline and allow light into the loft area of the barn. Their two important functions are ventilation, which helps guard against spontaneous combustion, and access for loading the hayloft. Gabled dormers are more common than shed dormers. Shed-roof dormers are earlier in design than gable-roof dormers, which did not appear generally before the latter part of the nineteenth century.

Fig. 3.26 The metal ridge ventilator which became popular in the twentieth century, especially in livestock and dairying areas.

BARN BRIDGES AND ENTRY PORCHES

BARN BRIDGES

Found on Bank and Raised barns, barn bridges provide access via a ramp to the upper level. They are usually found on the barn side, but sometimes on the gable (see Welsh barns). The bridge also may provide shelter for a tractor or other equipment. The ramp is usually built of earth and sometimes is hollowed out for supplemental use as a food cellar.

Range: Entire northeastern quarter of the United States and southern Ontario.

ENTRY PORCHES

An entry porch consists of a gabled roof and two side walls, and normally is located on the side of the barn. This feature probably has English origins. On larger barns these porches may appear in pairs on the same side of the barn.

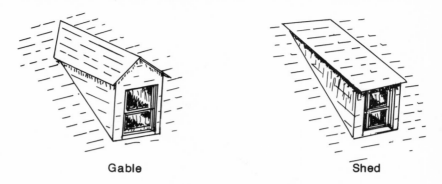

Gable   Shed

Fig. 3.27 Examples of gable and shed dormers. The shed dormer is earlier in design.

Range: From western New York to eastern Ohio, and lower Mohawk and Schoharie valleys of New York.

Variant. The entry porch is combined with the barn bridge or ramp. This structural design is probably a direct import from England.

Range: Northeastern Vermont; scattered across New York, northwestern Pennsylvania.

Fig. 3.28 A typical barn bridge gives access to this Raised barn. The bridge sometimes provides modest shelter for wagons or equipment. (Courtesy of Pioneer America)

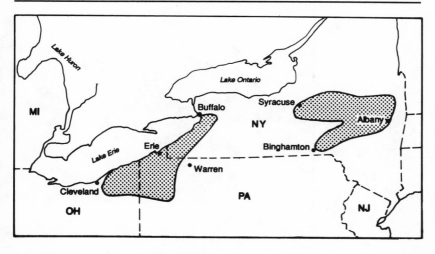

Fig. 3.29 Areas where barns commonly have entry porches. Of course not all barns within these areas have such porches.

WINDOWS, WALL OPENINGS, AND DOORS

Most early barns, and all simple barns, have few if any openings or windows, but each opening has an important function. Doors provide access, louvers and other openings ensure necessary ventilation, and windows give light.

Fig. 3.30 The entry porch is frequently combined with the barn bridge in certain areas. (Courtesy of Pioneer America)

Fig. 3.31 Rows of small windows providing light to this barn near Zee-land, Michigan, suggest it is used for dairying.

WINDOWS

One of the hallmarks of the more recent barns, such as the Erie Shore and Wisconsin Dairy barns, is the abundance of regularly spaced windows. The desirable objective of providing natural light could not be achieved until after the middle of the nineteenth century with the introduction of glass rolling mills, which reduced the cost of window glass.

The haphazard pattern of windows of different types in many barns indicates that farmers have added windows at different times to provide light. Windows or openings in the gable are common on barns that have major doors on the barn side. While many barns do not have openings high in the gable end, those that do fall into several patterns:

1. A simple rectangular window or opening is very common. On barns whose origins can be traced to the North German Plain, two additional windows are added lower in the gable.
2. A pair of rectangular windows is less common.
3. A rectangle tilted to parallel the line of the gable is frequently found, especially in the Northeast. Usually this opening is square, so it appears as a diamond shape. Most often it has four panes.

Fig. 3.32 A small square or rectangular window or opening high up in the gable is typical of many side-entry barns. This barn is in Geauga County, Ohio. Note also the rectangular silo.

       Range: Especially prevalent in Vermont, upstate New York, southern Michigan, Wisconsin, and southern Ontario.

4. Round windows or portholes are common in some areas and are sometimes presented as a pair or triplet.
5. Half-round windows appear in some locations and mirror a popular stylistic detail of early nineteenth-century houses. Many such windows appear on barns along Wisconsin's Lake Michigan shoreline north of Sheboygan. These windows are more likely to appear on carriage houses than barns.
6. A particularly ornate window pattern is typical of some Round-Roof barns.
7. A round window surrounded by a five-pointed star is typical of barns in the Belgian areas of Door County, Wisconsin.

WALL OPENINGS

Many gable-end windows are not glassed, but are simply ventilator openings, often with narrow, horizontal louvers. Barns

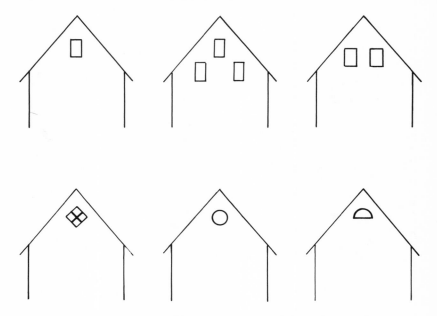

Fig. 3.33 Examples of typical gable-end windows or openings. These are only the most common; other types can often be found.

used for hay storage are especially prone to spontaneous combustion and need adequate ventilation. German barns are noted for arched ventilators of "Gothic" appearance.

OWL HOLES
Another feature to help ventilation and to allow entry of mice-eating owls, owl holes are always in the gable wall. They are usually found in multiples and may be shaped like diamonds, clubs, hearts, spades, or some other geometric figure. Owl holes are associated with both English and German settlement areas.

Other terms: Martin holes, swallow holes.

Range: Appalachia, from Pennsylvania to Ohio and south to Virginia; Wisconsin; southern Ontario.

SLIT VENTILATORS
Early stone barns often have long, narrow, slit ventilators sometimes mistakenly taken by romantics to be gun ports. Slit ventilators occur most often in Pennsylvania and surrounding

Fig. 3.34 Louver ventilators on a German Bank barn. Note also the two prominent ridge ventilators which emphasize the Victorian Gothic design. (Courtesy of M. Margaret Geib and the University of Massachusetts Press)

areas. Pennsylvania German barns (see chap. 4) of brick construction often have decorative brickwork in the gable end which serves as a ventilator. (See also figs. 1.5 and 1.6.) These barns are found mostly in south-central Pennsylvania and adjacent Maryland.

From the authors' experiences, patterns of windows and ventilation tend to be localized across relatively small areas. As with dormers, it is unclear whether these patterns represent a stylistic detail that has locally diffused, the mark of a particular barn builder, an ethnic trait, or all of the above.

DOORS

Doors are remarkably similar across the entire range of barns. In southern areas with milder weather, doors are frequently left off completely. Usually doors are rectangular and of two sizes: (1) large, double wagon doors which admit animals, wagons, and equipment; and (2) smaller, rectangular doors intended for humans. Early doors were hinged, but after the 1880s, sliding doors were used for most larger openings.

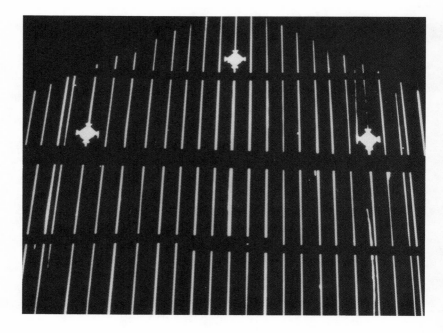

Fig. 3.35 Owl holes on a barn gable. These particular ones occur in a Standard Pennsylvania barn near Salem, Ohio. The view is from inside the barn.

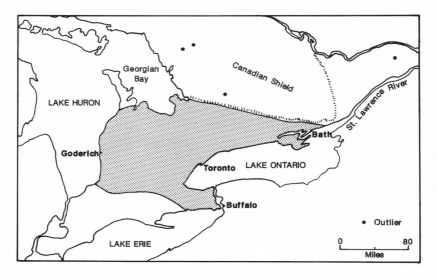

Fig. 3.36 The area of Ontario in which diamond-cross owl holes occur. (Courtesy of Pioneer America)

Fig. 3.37 Narrow-slit ventilators are often associated with German Bank barns. These are on a barn in Berks County, Pennsylvania.

Fig. 3.38 Sliding doors on a barn in Albany County, New York. Such doors became popular in the late nineteenth century to reduce damage from high winds when opened.

Variant 1. Door-in-door. A human-scale door is built into one of a pair of large wagon doors. This occurs only on swinging doors, not on sliding doors, and eliminates the need to open the unwieldy wagon door just to admit a person. Large doors are also more apt to be caught by wind gusts and to suffer damage to hinges.
Range: Widespread across eastern half of United States and Canada.

Variant 2. Dutch door. The top and bottom open independently, so the top can be opened for light and ventilation at the same time the bottom remains closed to keep animals in or out. This door occurs commonly on the lower levels of Bank barns.
Range: Wherever Pennsylvania German barns are found; also in New York.

Variant 3. Doors on Quebec Long barns sometimes have Norman curved tops rather than rectangular ones. Also found on some German barns, especially if built in stone.
Range: French Canada; Chester County, Pennsylvania; Gasconade County, Missouri.

Fig. 3.39 *The human-scale door-in-door is painted white to contrast with the larger red wagon doors on this out-shed barn in York County, Pennsylvania.*

Fig. 3.40 *Drawing of a barn with Dutch doors, whose top and bottom swing independently. This arrangement provides both ventilation and light but restricts animal movement.*

Fig. 3.41 *Norman doors on a Pennsylvania German barn are a direct transfer from the Old World.*

STRAW SHEDS AND HORSEPOWER SHEDS

STRAW SHEDS

In the nineteenth century, straw, especially that obtained from threshing rye, became a steadily more important farm by-product used for animal bedding, manure treatment, and commercial packing material. To protect the straw, which formerly had merely been thrown into a pile in the farmyard, rectangular extensions to the barn *rear* side were built. Eventually these extensions were incorporated into the design of new structures and a new barn type was born, the Three-End barn. Straw sheds are often two-story structures, attached to the upper level of the barn and extending to the ground. Sometimes the lower half of the shed is left open on one side and is used for machine storage.

HORSEPOWER SHEDS

A basically rectangular projection from the *front* of a Three-Bay Threshing, or Bank, barn, a horsepower shed is usually identifiable by cutoff wall corners and a pair of double doors

Fig. 3.42 An incipient straw shed raised on posts. The barn is located in Geauga County, Ohio.

opening at right angles, but toward the barn's main doors. This shed housed a device, propelled by a horse, that provided power for threshing.

Other term: Horse engine houses.

Range: Quite rare, but seen now and then in New York and Pennsylvania.

Fig. 3.43 The horsepower shed.

Fig. 3.44 Decorated barn doors typical of western Ohio. The barn is located in Wood County, Ohio.

BARN DECORATION

DOOR DECORATION

In some areas, door frames are painted a different color (usually white) from the barn proper (usually red). Fancy arches of several shapes frequently occur. Window and louver frames are also highlighted as well as corner posts.

Range: East-central New York, western Ohio, northern Indiana, southern Michigan; Wisconsin.

Variant. In the lower Schoharie valley, doors are often painted to provide two sets of contrasting triangles.

BARN MURALS

Scenes can range from elaborate landscapes to simple sunbursts. Painted advertisements are more common.

Range: Widespread, but infrequent; concentration in southern Wisconsin; Michigan, between Detroit and Flint; and a few in northern Vermont and New Hampshire.

DECORATIVE ROOFS

In the late nineteenth and early twentieth centuries, thin blocks of slate were often used for roofing materials. Because

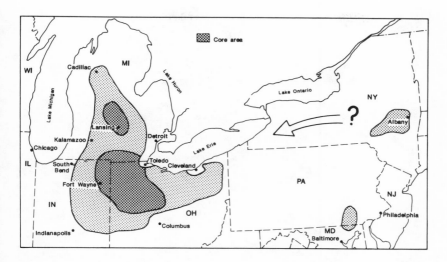

Fig. 3.45 Areas where barns commonly have decorated doors. A route of diffusion has been suggested but definitive conclusions cannot yet be drawn.

Fig. 3.46 Doors decorated with opposing triangles, Schoharie valley, New York.

Fig. 3.47 A humanized sunburst on a barn gable in Medina County, Ohio. Other barn murals may be much more elaborate.

Fig. 3.48 This decorated slate roof carries the name of the original barn owner. Much more typical is a slate roof with a date. This barn is in southern Lorain County, Ohio.

Fig. 3.49 A roof design in asphalt shingles, Fulton County, Indiana.

slate came in different colors, various geometric designs, dates, and names were often incorporated in the roof by careful placement of different colors.

Range: A few in eastern New York, Vermont, and eastern Pennsylvania; common in the eastern Midwest.

Variant. In the mid-twentieth century, asphalt composition shingles have supplemented slate. Some roofers have designed such roofs with pictures of farm scenes as well as owners' names.

Range: Widely scattered, but especially in the Midwest. Particularly commonly around Canton, Ohio, and Rochester, Indiana.

HEX SIGNS
The painting of hex signs on barns is an outgrowth of the earlier German tradition of decorating both barns and houses. The first hex signs in North America were geometric designs enclosed in circles painted on barns in southeastern Pennsylvania. These earliest signs were merely decorative, but gradually a folk tradition developed that these signs were symbols to ward off evil or to ensure good fortune. They have, however, entered the cultural tradition of Pennsylvania barn builders, even when that culture spread outside of Pennsylvania (to eastern Ohio, for example). They are most commonly encountered today on suburban garden sheds and hobby farms.

4

Crib Barns

The simple structures known as crib barns are essentially formed of a pen or crib of logs held together at the corners by notches. Sometimes the logs are covered by siding (usually vertical), but in more remote areas the logs are always exposed. Crib barns may have a German genesis, although Double-Crib barns may well be Finnish in origin. Regardless of the origins of the type, Scotch-Irish and English settlers also used the structure to advance settlement in North America. Later on, Finnish and Scandinavian settlers introduced crib barns directly into the northern Great Lakes region. (See fig. 2.2 for evolution of floor plans.) Crib barns are found throughout Appalachia, but are uncommon in the northern part. They are also found in southern Ontario and eastern Michigan as well as the northern Great Lakes area.

SINGLE-CRIB BARN

Usually 8–12 feet on a side, a Single-Crib barn is probably the simplest barn—a crib or pen of rough-hewn logs and simple gable roof. The logs are not chinked, so the crib is not weathertight unless boarded over. The main door may be on the side or gable end. The gable end is usually the narrower of the two dimensions. Many barns have a small opening for grain loading into the loft, which is tightly enclosed by sawn boards.

Range: Hill country of Appalachia, especially southern part; Maryland and coastal Virginia westward through Kentucky and Tennessee to the Mississippi River and less commonly throughout the Deep South. Single-Cribs also found on the Canadian Shield, in the Ozarks, and in isolated occurrences throughout the mountainous West.

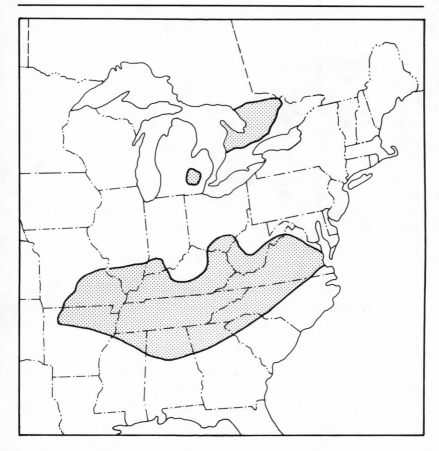

Fig. 4.1 Area where crib barns are most common in eastern North America. The largest area is in southern Appalachia.

Variant. A Single-Crib barn with flanking sheds has a lean-to shed on one or both sides. The side sheds are usually of lighter, sawn lumber with a lower roof pitch than the central crib, often indicating their addition at a date later than the construction of the central crib. The main door is on the gable end. These expanded Single-Crib barns are more common to-day than are Single-Crib barns.

DOUBLE-CRIB BARN
The cribs are of rough logs, although some Double-Crib barns may have hewn-timber or even sawn-lumber siding. Sometimes the two cribs may be of different dimensions. Both cribs

Fig. 4.2 Single-Crib barn, Calhoun County, Illinois. (Photo by H. Wayne Price)

share a common roof extended across a central aisle, open space, or threshing platform. Note the ridgeline across the cribs (transverse) and the door placement, which differentiate the following types.

TYPE 1

The longer dimension parallels the ridgeline, and crib doors open onto the aisle. Used for stabling in the Upland South and corn storage in the Deep South. Cribs are smaller as one goes farther south.

TYPE 2

The cribs are more nearly square, but the shorter dimension parallels the ridgeline and the crib doors open to the barn side rather than to the central aisle. Sometimes a pent roof over the doors is supported by gable log extensions. It has a possible connection to the English barn.

Variant. A mid-Atlantic variant has a frame shed extension to one side and a forebay extension on the opposite side. This may be a derivative of German "ground barns" or Grundscheier.

TYPE 3

Composite of Types 1 and 2. The shorter dimension parallels the ridgeline, but the crib doors open to the aisle. Cribs are divided by lighter material, with a central partition into two spaces on each side of the central aisle. Not as common as other types.

Fig. 4.3 Double-Crib log barn from Pike County, Ohio.

TYPE 4

Unequal-size cribs. The larger crib's long dimension is parallel to the ridge, but the smaller crib is transverse. Generally, crib doors open to the aisle, although some are known to open directly to the outside. Typically the larger crib measures 16

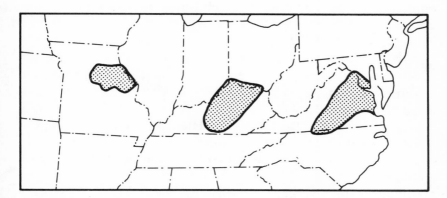

Fig. 4.4 Probable areas of Type 4 Double-Crib barns. Such barns are reported in Missouri and adjacent Calhoun County, Illinois. Central Kentucky and eastern Virginia are the source areas of settlers in Missouri and thus are likely to have the same barns.

Fig. 4.5 A Cantilevered Double-Crib barn located in Cades Cove, Tennessee. This is perhaps the most beautiful barn type in North America, and this particular barn may be the most photographed barn in the United States.

feet by 16–21 feet, while the smaller crib is 16 feet by 9–11 feet, often with vertical slats for corn storage.
Range: Calhoun and Hardin counties, Illinois, Little Dixie region of Missouri, and perhaps elsewhere in early corn-growing areas such as the Shawnee Hills of southern Illinois.

CANTILEVERED DOUBLE-CRIB BARN
In this barn the loft may overhang cribs on all sides, although loft overhang on only the front and back is more often encountered (similar to German forebay construction in southeastern Pennsylvania). The large loft is used for hay storage, and doors open to the barn side. The cribs are always log, but the upper portion is as likely to be frame as log. Because of its distinctive appearance, this barn will not be confused with any other kind.
Range: Eastern Tennessee, western North Carolina, eastern Kentucky, border of Virginia and West Virginia.

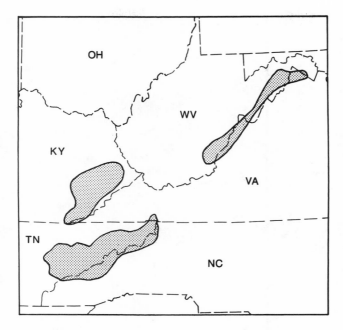

Fig. 4.6 Areas where Cantilevered Double-Crib barns have been reported.

Fig. 4.7 A Front-Drive Crib barn from Jackson County, Kentucky.

Fig. 4.8 A Side-Drive Crib barn with a shed added on to the crib side, near Gimsley, Tennessee.

FRONT-DRIVE CRIB BARN
The crib is approximately 8 by 8 feet, the roof approximately 8 by 16 feet. This barn is a single crib with a forward projecting front roof supported by corner poles. The loft may be enclosed for grain storage. The ridge runs from front to back.
Range: South and central Appalachia.

SIDE-DRIVE CRIB BARN
The approximate dimensions overall are 12–20 feet roughly square. The aisle and crib are about equal dimensions, each covered by about one-half the roof, which has unbroken slopes. The ridge runs from front to back. Sometimes a shed is added to the side away from the aisle. The loft siding is more weatherproof than that of the crib. The aisle is sometimes enclosed.
Range: Widely scattered across Kentucky, probably elsewhere in Appalachia.

Fig. 4.9 A Drive-in corncrib in Tazewell County, Virginia.

Similar barns: Single-Crib with one flanking shed, but this usually has a definite break in the roofline between crib and shed, indicating that the aisle was a later addition.

DRIVE-IN CRIB BARN
This barn has elongated cribs. The loft is small or absent, and sometimes it has roof hatches. Usually the barn is of frame construction. The driveway is without doors, front and back. This barn is always used for corn storage. The aisle may be used for equipment storage.
Range: Wide distribution from Virginia to Kansas and Nebraska, north at least to Ohio and western Pennsylvania and New York.
Similar barns: Double-Crib barn, but ridgeline is transverse. See also "Corncribs" in chapter 12.

FOUR-CRIB BARN
The approximate dimensions are 24 by 40 feet overall. Four rough-hewn log cribs are located one at each corner, with a single gable roof covering entire structure. The cribs are usu-

Fig. 4.10 The Four-Crib log barn. (Courtesy of M. Margaret Geib and the University of Massachusetts Press)

ally of similar size (8–16 feet). Few of these barns exist, and no frame examples are known. Side-aisle openings are often filled in by sawn siding, so the structure may appear to be a Transverse Frame barn.

Range: North-central Tennessee, south-central Kentucky (Cumberland valley).

5

Crib-Derived Barns

Crib-derived barns are barns that, by process of elaboration, have been derived from the earlier and simpler crib barns (see fig. 2.2 for floor plans). They are universally of frame construction, although log pens or cribs may continue to be present within the structure, often functioning as a basic unit around which the larger, more elaborate structure has been designed. These crib-derived barns are most common in Appalachia and the eastern and northern Midwest.

APPALACHIAN BARNS

This is a full barn rather than a roofed crib, so it is much larger and of frame construction. Aisles are often fully or partially enclosed. Usually no main door opens off the gable end, but there is a large loft opening. Often this barn has a hay hood (see fig. 3.23). It may be related to the Gable-Forebay German barn from southeastern Pennsylvania.

Range: Common in southern and central Appalachia and parts of Missouri and Arkansas, and as far north as southern Ohio and Indiana.

TYPE 1

Two or three internal cribs with a cross aisle in front. Crib doors open onto the aisle. Frequently the side of the aisle is left open. Many floor plan variations.

TYPE 2

A short transverse aisle between two cribs forms a T with the main front aisle. Crib doors open onto a transverse aisle. The front gable-end is sometimes expanded to form a narrow corncrib. A hay hood is common.

Fig. 5.1 An example of an Appalachian barn, Type 1, located near Lexington, Missouri.

TRANSVERSE FRAME BARN

This barn may be derived from the Four-Crib barn but has an all-frame construction. More likely it was introduced directly from northern Germany. A wagon door is in one or both gable ends.

TYPE 1

Most are slightly longer than wide, with gable walls 24–30 feet and side walls 28–36 feet. Doors are in the gable end; the main aisle is centered. Often there is no loft, so the entire interior is open. Frequently, but not always, used as a tobacco-drying barn.

Range: Virginia through Illinois, western Wisconsin.

Variant 1. In southern Ohio and Indiana, a smaller, shorter Transverse Frame barn is common.

Variant 2. In the Bluegrass Basin, the U.S. Department of Agriculture promoted a standardized barn design of lower roof pitch and greater length than the basic Transverse Frame barn.

Range: Bluegrass Basin of Kentucky.

Fig. 5.2 An Appalachian barn, Type 2, from near Cades Cove, Tennessee.

Fig. 5.3 A Transverse Frame barn located in Shelby County, Kentucky. Note doors on the gable end. The ventilators along the ridge indicate that this barn's function is to dry tobacco.

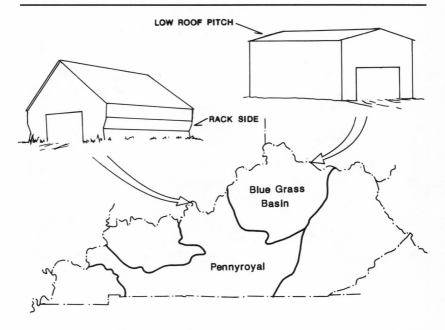

Fig. 5.4 Bluegrass Basin Transverse-Frame tobacco barns and Pennyroyal rack-sided livestock barns show the differences in form which are possible over relatively short distances.

TYPE 2

A rack-sided barn, with slanted lower sides, interior mangers, and a central aisle.

Range: Pennyroyal area of central Kentucky; isolated examples elsewhere (Sevier County, Tennessee, for example).

MIDWEST THREE-PORTAL BARN

This Transverse Frame barn has a central aisle and enclosed side aisles. Sometimes added aisles are later additions and often result in a broken roofline. Early gable roofs have sometimes been replaced with gambrel roofs, and in the twentieth century barns were built with original gambrel roofs spread to cover the side aisles. This barn was standard in the south-central United States in the late nineteenth and early twentieth centuries. It often has a hay hood and large gable-end loft doors. Plank frame versions are usually called Feeder barns as they house livestock. It may be related to barns derived from North German Plain.

Fig. 5.5 Midwest Three-Portal barn from Polk County, Iowa. The barn is timber-framed and each of the three gable doors leads to an aisle.

Other term: Three-alley barn.
Range: Scattered in western Kentucky, southern Indiana, and Illinois, and throughout the central Midwest.
Similar barns: Single-Crib barn with flanking sheds, Transverse Frame barn.

Variant 1. The central aisle is large with wagon doors at gable end. Gable roof.

Variant 2. Same as variant 1, but the broken roofline suggests later addition of flanking aisles and sheds.

Variant 3. The central aisle is reduced to a narrow walkway with flanking cribs or stables added. The gable wall often is longer than the side wall. The side measures 36–42 feet. Gable roof. Roofline sometimes broken.

Other variants. Same as variants 1, 2, and 3, but with a gambrel roof instead of a gable roof. Generally of smaller size, reflecting the gasoline engine's ascendancy over the horse.

6

Ethnic Barns

More than half the barns in this book have some relation to one of the ethnic groups that settled North America. Sometimes that relationship seems trivial, as in the five-pointed star that Belgian settlers often affixed to the gables of the Raised, English, Erie Shore, and Dairy barns that they adopted from their neighbors in their new homes. In some cases, the process of transplanting an Old World barn onto New World soil has been revealed by painstaking research. English barns in New York look much like English barns in England and are most prevalent in areas of early English settlement. Swedish barns in parts of Minnesota and Wisconsin look just like barns in Sweden, but many other Swedish settlements in North America seem to have abandoned their older forms and instead borrowed barns from their neighbors. In too many cases, however, the process still remains mysterious. Cajun barns in Louisiana were clearly built by Acadians, but there seems to be no prototype for this barn in either Nova Scotia or France.

A 1989 award-winning book by Terry Jordan and Matti Kaups, *The American Backwoods Frontier*, has linked the Double-Crib barn to Finnish antecedents and has provided a context for this diffusion, one that threatens our long-held ideas of North American frontier settlement. A more recent groundbreaking work by Robert Ensminger, *The Pennsylvania Barn* (1992), reveals that the early Pennsylvania German Sweitzer barns are virtually identical to a barn type in Switzerland. Perhaps more importantly, he traces the evolution of that barn type into some distinctive North American forms never seen in the old country.

Most North American barns are linked in one way or another with emigrant ethnic groups. The abandonment of ancestral forms, the transplantation of old types into a new environment, and the adoption of types already existing in a new area are clues to cultural change, but unresolved questions abound.

ENGLISH BARNS

The English barn in America clearly follows a traditional English design. In England these barns are almost always built of stone, while in North America the building material is almost always timber. To an extent, this situation reflects the lack of suitable timber in England and its abundance in the New World. Despite these differences in building materials, the form of the structures is identical.

The later New England Connected barn is simply an English barn connected to the farmhouse by a series of intervening frame buildings. While no direct English antecedents of this barn appear to exist, English barns under the same roof as the house, or connected to the house and other buildings around a courtyard, may have inspired this type. The Gable-Entry Bank barn, the English Bank barn, and the Raised barn may have antecedents in Cumbria and the English Lake District, although a possible Scots and Devonshire connection also exists for the latter, and a possible Welsh origin exists for the former. The Foundation barn appears to be a wholly North American derivitive, however.

THREE-BAY THRESHING BARN
This barn is timber-framed, with post-and-beam construction. It is side-gabled with a central runway and usually equal-sized spaces on either side (e.g., three bays including the runway). The bays are sometimes divided by lightweight construction. The barn has a single story, with a loft for hay storage. Early examples have steeply pitched (over 45 degrees) gable roofs. There is a low stone foundation, or sometimes the barn sits on individual large rocks. It is sided with vertical boards, and small ventilation openings are high on the gable ends. Windows are largely absent. Gable-end sheds are a common addition in the Midwest.

Fig. 6.1 An English or Three-Bay Threshing barn from northwestern Ohio. Note the prominent gable-end pent roof.

Other terms: English barn, New England barn, Yankee barn, Three-Bay barn, Connecticut barn.

Range: New England, New York, eastern Midwest through lower Michigan and eastern Wisconsin.

Similar barns: Grundscheier.

Variant 1. Some Three-Bay Threshing barns are constructed with a swing beam, a beam larger than the others which extends unsupported from front to back of the barn on one side of the threshing floor. It permits threshing of grain and turning of teams, unhindered by a post.

Variant 2. Unhewn, notched log barns of the English type with plank roofs and gables are known from Mormon Utah and also from northern Wisconsin.

Variant 3. English barn with gable-end shed. Common in Midwest.

Variant 4. English barn with barn-length shed across rear side. The shed creates a saltbox-roof appearance. Common in northeastern Wisconsin.

Fig. 6.2 Log version of the English barn from Sevier County, Utah. (Courtesy of M. Margaret Geib and the University of Massachusetts Press)

NEW ENGLAND CONNECTED BARN

Not truly a separate barn type, this structure is often an English barn connected to the farmhouse via a series of intervening structures, usually *en echelon*. Each building has a separate roofline. The particular form varies widely, but parts are always joined. Usually it has four major parts, which show up in the first line of the familiar New England nursery rhyme, "Big house, little house, back house, barn."

Range: New England and extreme eastern New York. (See fig. 2.1 for geographical distribution.)

ENGLISH BANK BARN

Entry to this barn is on two opposing levels due to sloping ground. The partially excavated lower story is used primarily to house animals. The upper part of the structure is identical to the Three-Bay Threshing barn from which it is probably derived.

Other term: Side-Hill barn.

Range: New York State, especially eastern part; western New England, northeastern Ohio.

Fig. 6.3 The New England Connected barn.

Similiar barns: German immigrants, especially in Pennsylvania, also built bank barns, but these are characterized by a forebay.

Variant. Stone bank barn with ridgeline parallel to the slope. Ventilator slits. Two pent roofs and a series of single doors (normally five) on downslope side. (Identified by

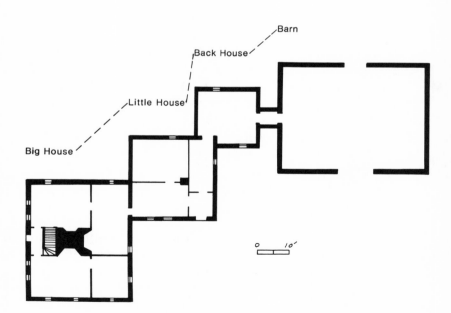

Fig. 6.4 Floor plan of a New England Connected barn keyed to the popular childhood rhyme. (Courtesy of M. Margaret Geib and the University of Massachusetts Press)

Fig. 6.5 The English Bank barn or Side Hill barn.

Charles Dornbusch and J. K. Heyl, in *Pennsylvania German Barns*, as Type E.)
Similar barns: Raised barn, attached-forebay, Double-Decker.

RAISED BARN

Commonly 30–50 feet wide by 60–100 feet long, this English barn is raised on a stone, brick, or concrete foundation. Doors to the basement are usually on the gable ends, but sometimes on the downslope side. Often it has inaccessible threshing doors opening to the rear side. Usually it rests on level ground

Fig. 6.6 English Bank barn variant (Dornbush and Heyl Type E).

Fig. 6.7 The Raised barn. (Courtesy of Pioneer America)

with an earthen ramp or barn bridge built to the wagon doors. Sometimes it has a covered-roof entry structure. Expanded versions with up to five bays exist.

Other terms: Raised three-bay barn, basement barn, bank barn, central or southern Ontario barn.

Range: Upper Midwest, New York, Vermont, southern Ontario.

Similar barns: English Bank barn usually on a slope with no gable-end doors. Sweitzer and other German Bank barns, but these have an overhanging second story or forebay.

Variant. In the panhandle of western Pennsylvania, extreme southwestern New York, and northeastern Ohio, Raised barns often have large entry structures.

FOUNDATION BARN

Early versions, especially, appear identical to Raised barns, with one-story, fieldstone foundations, no windows, gable roofs, and vertical cladding. However, these barns have no entry to the second level via ramp or bank. These barns typically have ridge pole extensions or hay hoods and probably represent an adaptation of the Raised barn to modernizations in hay-loading equipment. The main wagon door is usually on one gable end. More recent barns have concrete foundations and often have many windows.

Range: This recently identified barn is common in central and northern Wisconsin; central New York.

Similar barns: Raised barns have ramps or are built into a bank.

Fig. 6.8 The Foundation barn. There is no ramp, bridge, or other access to the upper part of the barn from outside.

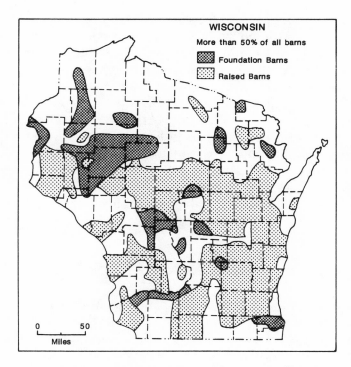

Fig. 6.9 Distribution of Raised and Foundation barns in Wisconsin.

Fig. 6.10 A Gable-Entry Bank barn located in Oneida County, New York. Note the long bridge to the upper floor.

WELSH GABLE-ENTRY BANK BARN

This barn is situated with the ridgeline perpendicular to the hill slope. Upslope entry is through the gable end; basement entry is through the opposite gable. There are two floors plus loft, with a gable roof. Found in areas of Welsh settlement, but may have origins in English Lake District.

Range: Western New England, especially Vermont; central New York, north-central Pennsylvania, especially in Bradford County around Neath; Wisconsin.

Similar barns: Grundscheier (Dornbusch and Heyl, Type C), but with differences in location of doors.

Variant. Single-story, gable-entry barns occur in central-western Ohio in Welsh-settled areas.

GERMAN BARNS

German barns are perhaps the most common barns found from eastern Pennsylvania to the central Midwest. They were introduced into southeastern Pennsylvania at the end of the seventeenth century. Many of these barns are similar to hillside barns found in parts of Europe, so considerable debate

Fig. 6.11 A one-story Welsh barn in the settlement of Gomer, Ohio.

revolves around the degree to which the designs evolved in the New World.

Until recently there has been considerable confusion in the terminology associated with these barns. Many are bank barns, that is, built into a bank or with a ramp to the second level, and many are forebay barns, with an overhanging second level on one or more sides. Almost all forebay barns are banked, but not vice versa. The Pennsylvania barn, a new classification that is the outcome of research by Robert Ensminger (see source list), is a broad class of these banked forebay barns.

German barns must be viewed from both upslope and downslope in order to best see many of their distinguishing characteristics. Their classification is more complex, and important differentiating details are better known than those of other barns, because German barns have been more carefully studied. For almost every type, however, the barn can be identified correctly from an exterior view.

GRUNDSCHEIER

A "ground barn" of German origin, sometimes it is situated on a level site, but more often on slightly sloping ground.

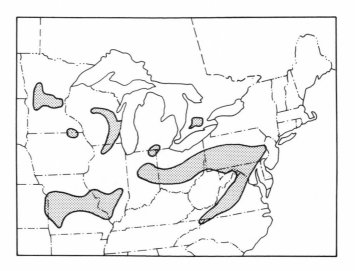

Fig. 6.12 Areas in which Pennsylvania (German) barns are most commonly found.

Fig. 6.13 A stone Grundscheier located near Oley, Pennsylvania. The stable part is to the left.

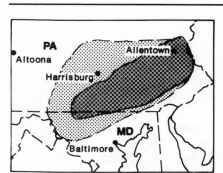

Fig. 6.14 Areas in southern Pennsylvania and adjacent Maryland in which Grundscheier can be found. The darker section represents the core area where the largest number and the earliest examples of this barn occur.

Early forms may have two entry levels (because of slope), but most examples have three separate ground levels. The lowest level is the stable with haymow above. The ridge is perpendicular to the slope. The lowest level may be partially excavated and extends only about a third of the length of the barn. Grundscheier are no longer very common.

Other terms: Tri-level ground barn.

Range: Southeastern Pennsylvania, especially Bucks County.

Similar barns: English barn, Double-Crib barn Type 2.

GRUNDSCHEIER WITH FOREBAY

This single-level barn has two log cribs separated by a threshing floor. The framed forebay is usually split in two by threshing doors.

Range: Bedford, York, and Adams counties, Pennsylvania; from Virginia/West Virginia border to eastern Kentucky, eastern Tennessee, and western North Carolina.

Similar barn: Double-Crib barn.

Variant 1. The framed forebay extends unbroken across the entire length of the barn so that it looks like a small Sweitzer barn (see below). (See fig. 2.3 for an example of this barn.)

Other terms: Half Sweitzer.

Range: Central Maryland, southeastern Wisconsin; rare in eastern Ohio.

Variant 2. The forebay is of frame construction and appears on one gable end of the barn; very rare.

Range: Oley valley of Pennsylvania.

Fig. 6.15 A Grundscheier with a split forebay located in Lebanon County, Pennsylvania.

Fig. 6.16 A Grundscheier with a gable forebay. Located in the Oley valley, Pennsylvania.

Fig. 6.17 Grundscheier barn Type D (Dornbush and Heyl).

GRUNDSCHEIER TYPE D

This barn is large, unbanked (level ground), with three bays. All-stone or partial stone walls are of varying height. Often it has ventilator slits rather than windows. There is a slight ramp to the wagon doors, which are very tall, usually extending to the eaves. Sometimes a pair of pent roofs are on either side of wagon doors. First identified by Dornbusch and Heyl (*Pennsylvania German Barns*), this barn may be English rather than German in origin.

Range: Southeastern Pennsylvania, especially Berks and Bucks counties, southern Wisconsin.

Similar barns: English barn.

CLASSIC SWEITZER BARN

This barn is identified by its unsupported overhang on the downslope side of the barn and its asymmetrical roofline when viewed from the gable. The roof pitch is steep (40–45 degrees), and the ridgeline is parallel to the hill slope. Typically it is three bays wide by two bays, plus forebay, deep. The lower story is usually stone, although stone may have been replaced by concrete. The lower story is partially excavated in the bank of a hill. Double wagon doors are on the upslope side. Inaccessible (from the exterior) threshing doors are above the stable yard (feedlot).

The most prominent feature is the overhanging forebay supported by cantilevered beams over the feedlot. While the barn may be log, half-timbered, timber frame, brick or stone,

Fig. 6.18 A Sweitzer barn located in Ashland County, Ohio. Note the asymmetrical roof profile. The door in the forebay is not always present.

the forebay is always timber frame with plank siding. Often a series of Dutch doors under the forebay open into the basement level.

Other terms: Schweitzer, Swisser (Dornbusch and Heyl Type F, Type G).

Range: Common from Pennsylvania to central Midwest, Shenandoah valley of Virginia, central southern Ontario; almost always an indicator of original German settlement.

Variant 1. Log Sweitzer. Typically about 60 by 30 feet, with a forebay overhang of 6 feet. Structure consists of two log-crib mows on either side of a threshing floor. Frame forebay. Stone basement usually. No ridge board.

Range: From southeastern Pennsylvania to Shenandoah valley, westward to Ohio, southern Ontario.

Variant 2. Transitional Sweitzer. Same as other Sweitzers, except basement wall is extended to enclose the forebay.

Range: Southeastern Pennsylvania, especially Lancaster County.

STANDARD PENNSYLVANIA BARN

This barn differs from the Sweitzer barn in that it has a symmetrical gable profile. The forebay is incorporated into the

Fig. 6.19 A Transitional Sweitzer barn from near Hamburg, Pennsylvania. The notched stone wall is an important diagnostic feature.

main barn framing and is generally shallower than that of the Sweitzer barn. The roof pitch is 30–35 degrees. Granaries are in the forebay. The following five barn types and four special forms are all major variants of the Standard Pennsylvania barn.

Range: Pennsylvania to Wisconsin, Virginia, southern Ontario; infrequent in Missouri, southeastern Iowa, and Erath and Bell counties, Texas.

CLOSED-FOREBAY STANDARD BARN

This structure is one of the major subtypes of the Standard Pennsylvania barn. Stone gable ends enclose a frame forebay less than 6 feet deep. The upslope sidewall is also sometimes framed. The area under the forebay appears recessed.

Other term: Dornbusch and Heyl Type H.

Range: Pennsylvania (especially Northampton, Montgomery, Lehigh, and Berks counties), central Maryland, extreme northern Virginia, extreme eastern West Virginia, eastern and central Ohio (entirely wooden construction), eastern Wisconsin (sometimes one-third to one-half of area under forebay occupied by a room with stone or sometimes wooden wall), a very few in Iowa.

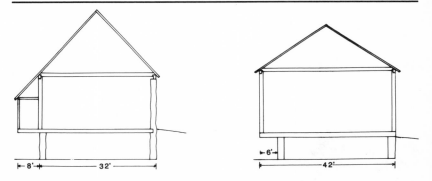

Fig. 6.20 A comparison of the Sweitzer and Standard Pennsylvania gable profiles. The Sweitzer is to the left.

Similar barn: Raised barn when viewed from upslope.

OPEN-FOREBAY STANDARD BARN

Usually this barn is characterized by a fieldstone foundation, frame upper stories, and a forebay incorporated within the barn-framing system; it is cantilevered over the downslope feedlot. Symmetrical gable profile.

Fig. 6.21 A Closed-Forebay Standard barn from near Allentown, Pennsylvania.

Fig. 6.22 An Open-Forebay Standard barn located in Portage County, Ohio.

Range: From central Pennsylvania across Ohio, Indiana, Illinois to Wisconsin, Iowa (scarce), and Nebraska (rare).

Similar barn: Should not be confused with Sweitzer barns, which have an asymmetrical gable profile and often a forebay not framed into the barn-framing system.

POSTED-FOREBAY STANDARD BARN

A series of wooden posts supports the wooden frame forebay, which may be up to 15 feet deep. Symmetrical gable profile. Open-forebay end walls. Some barns in Pennsylvania have half stone, gable walls.

Range: Pennsylvania, Shenandoah valley of Virginia, southern Ontario (especially Waterloo County), Ohio, Indiana, Illinois (especially Adams County), Wisconsin, a few in Iowa.

MULTIPLE-OVERHANG STANDARD BARN

The upper floors overhang the basement not only on the downslope, forebay side, but on one or more other walls as well. Gable-wall overhangs are narrow (1–2 feet). Upslope-wall overhangs are wider (2–6 feet), but rarely exceed the forebay overhang.

Range: Scattered across Pennsylvania and the Midwest.

Fig. 6.23 A Posted-Forebay Standard barn from Adams County, Pennsylvania. This particular barn has brick gable walls.

Important clusters include Shenandoah, Rockingham, and Augusta counties, Virginia; Perry, Fairfield, Allen, and Putnam counties, Ohio; Onondaga County (near Plainville), New York; Green County, Indiana

BASEMENT DRIVE-THROUGH BARN

A Standard Pennsylvania barn, but with a basement drive-through at one end of the structure. In some instances, the basement drive-through is accommodated in a one-bay extension of the main barn, so that the main bank entrance is off center.

Range: South-central Pennsylvania (especially Fulton and Franklin counties); central Maryland, Shenandoah valley of Virginia, scattered in Ohio and eastern Midwest.

SPECIAL FORMS OF THE STANDARD PENNSYLVANIA BARN

As might be expected, many specialized forms of the Standard Pennsylvania barn evolved. Most were built in limited numbers and hence are not often encountered today. Even the most intrepid barn hunter may never encounter all the types. The following four are the best known of these special forms:

Fig. 6.24 A Multiple-Overhang Standard barn located in Summit County, Ohio.

GABLE-RAMP BARN

Barn in which a gable ramp offers access to the second level. Not to be confused with the Welsh Gable-Entry barn, which has no forebay.

Range: A few in Berks County, Pennsylvania.

Fig. 6.25 The Basement Drive-Through barn.

GABLE-FOREBAY BARN
The forebay appears on the gable rather than on the barn side.
Range: A few in Berks and Montgomery counties, Pennsylvania; rarely found in Appalachia and other regions.

STONE-ARCH FOREBAY
The forebay wall is of stone, pierced by Romanesque arches (Dornbusch and Heyl Type K).
Range: Found mostly in southeastern Pennsylvania; Warren County, New Jersey; New Castle County, Delaware.

BANK-INTO-FOREBAY BARN
The forebay is located on the upslope side of the barn rather than the downslope side. This necessitates entry from the bank directly into the forebay.
Range: Only four examples are known: two in Bucks County, and one in Centre County, Pennsylvania, and one in Morgan County, Missouri.

EXTENDED PENNSYLVANIA BARN
This banked stone, or stone and wood, forebay barn has a very large wooden forebay which appears to be attached to the barn rather than integrated with the rest of the structure. Usually it has a broken roof pitch, and the forebay is supported by large posts. The roof is asymmetrical. The following four barn types are the major variant forms.

EXTENDED SUPPORTED-FOREBAY BARN
The forebay, supported by large distinctively tapering stone piers, extends 15–20 feet from the barn main wall. A crossbeam connects the tops of the piers and forms the base of the forebay outer wall. Note the separate roofline of the forebay. This barn may represent a combination of German and English design elements.
Range: Chester and surrounding counties of southeastern Pennsylvania

UP-COUNTRY POSTED-FOREBAY BARN
The pronounced forebay is supported on wooden posts. Early examples were constructed of stone, later ones of frame. Often this barn has a gable dormer centered over the forebay. It has a longer and generally lower silhouette than the Sweitzer barn. The forebay projects 9–12 feet and is always supported by timber posts. The posts are often surmounted by boards cut on a 45-degree angle, resulting in an arcaded appearance. Timber frame construction, usually five to seven

Fig. 6.26 The Extended Supported-Forebay barn.

bays wide, often with several pairs of threshing doors. Sometimes has wooden louvers, on both side and gable ends.

Other term: Wisconsin Porch barn.

Range: Southeastern Pennsylvania (especially Montgomery, Lehigh, and Berks counties); central Maryland and Virginia (especially Shenandoah valley), Ohio, Indiana, Illinois, Wisconsin (especially Lincoln, Marathon, and Ozaukee counties).

Variant 1. The area under the forebay is completely enclosed by a fieldstone foundation. Asymmetrical but unbroken roofline gives saltbox appearance. Clearly an Up-Country Posted-Forebay barn, but built as a single unit.

Range: Central and southeastern Wisconsin.

Variant 2. The area under the forebay is enclosed by a fieldstone foundation, or occasionally by an all-wood enclosure; the broken roofline creates the appearance of an "attached" unit. Internal examination reveals that these are sometimes Raised barns with attached, uncantilevered, and enclosed forebays.

Range: Concentration in central Wisconsin, but scattered through eastern and southern Wisconsin.

FRONTSHED EXTENDED PENNSYLVANIA BARN

An L-shaped barn resulting from the addition of a large gable-roofed structure with a forebay of its own to the downslope side of a forebay barn.

Fig. 6.27 An Up-Country Posted-Forebay barn located near Reading, Pennsylvania.

Other terms: Three-Gable barn.
Range: Southeastern Pennsylvania, central Ontario, northeastern Ohio.
Similar barns: Three-End barn, but it is ground level only with no forebay.

OUTSHED EXTENDED BARN

The asymmetrical gable profile is created by the addition of a 12–15 foot deep granary outshed(s) on the upslope side. The downslope side maintains the forebay. The outsheds often have an exterior door opening off the ramp and sometimes a door opening on the side away from the ramp as well.
Range: Southeastern Pennsylvania (especially Lancaster County); central Pennsylvania (especially Lycoming and Union counties); northern Maryland (especially Washington county); scattered locations in the eastern Midwest.'

Variant 1. Ramp-Shed Barn. The ramp area between the outsheds is walled in and roofed over, thereby extending the area and size of the barn.
Range: Pennsylvania; Ohio (especially Fulton County); Indiana (especially Adams County).

Variant 2. Decorated Brick-End Barn. This barn is a wedding of German barn construction and English decorative

Fig. 6.28 A Frontshed Extended Pennsylvania barn located near Moselem, Pennsylvania. The shed is to the left and the main barn to the right.

Fig. 6.29 An Outshed barn near Millersburg, Ohio. Note the asymmetrical roof profile whose longer side is away from the forebay.

Fig. 6.30 The Double-Decker German barn. The siding is an index to the number of stories.

brickwork. The gable walls consist of bricks laid so that decorative openings are formed by omitting certain bricks. The designs consist mostly of diamonds, hourglasses, double diamonds, wheat sheaves, dates, and initials (see figs. 1.5–1.7). Range: South-central Pennsylvania and north-central Maryland; rare in Ohio.

DOUBLE-DECKER BARN

This three-level bank barn plus a loft is usually built on a steep slope. Often entry to the threshing floor is via a covered barn bridge that protects a lower entry into the granary. Early versions are stone, later ones timber frame. Often a pent roof along the stable yard side.
Other terms: Dornbusch and Heyl Type L.
Range: York, Pennsylvania, area; German-settled Midwest.
Similar barns: Dornbusch and Heyl Type E is only two levels.

MADISON COUNTY AMISH BARN

An off-center wagon entry under a projecting gable is the distinguishing feature of this barn. A pent roof runs along the front of the barn instead of a forebay. The barn has Dutch doors, a right-angle straw or hay shed, and geometric cutouts high on the gable wall. Note the lack of lightning rods.
Range: Madison County, Ohio.

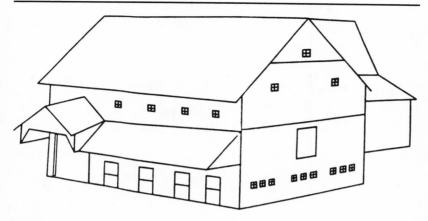

*Fig. 6.31 Amish barn, typical of those erected in Madison County, Ohio.
(Courtesy of the* Ohio Geographer: Recent Research Themes)

FRENCH BARNS

The barns erected by various French ethnic groups are not as widespread as those of the English or Germans. Hence they are not as well known and have not been studied as intensively. This is especially true about their links to European source areas.

QUEBEC LONG BARN
Lengths of 48–80 feet are not uncommon. This barn is side-gabled and timber-framed. Like the English barn but with elongated plan, the barn is often eight or more bays wide. An off-center dormer entrance leads to the loft. Often it has flared eaves. This barn is the French equivalent of the New England Connected barn, and both are partial responses to a severe winter climate.
Range: French Canada.
Similar barns: New England Connected barn, Czech barn.

MADAWASKA TWIN BARN
Two parallel, rectangular barns, usually identical in form and structure, are connected by a low, intermediate passageway.
Range: St. John valley of extreme northern Maine, southern Quebec.

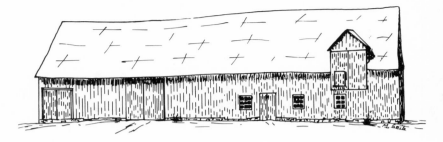

Fig. 6.32 The Quebec Long barn.

Similar barn: In western Ohio (Fulton and surrounding counties), twin barns occupy the northwestern corner. These double barns have no relation to the Madawaska Twin barn, although they have a similar form. In these barns, frequently the two parts were built at different times and often are not identical. For example, one may have a gable roof and the other a gambrel roof.

ACADIAN BARN

This barn consists of a rectangular, one-and-a-half story structure resembling an English barn, with a low extension covered by a lower-pitched shed roof. The result is a squarish floor plan. The higher part of the barn contains a haymow, threshing floor, and granaries. The extension contains animal stalls, stables, and drive floor.

Fig. 6.33 The Madawaska Twin barn. (Courtesy of M. Margaret Geib and the University of Massachusetts Press)

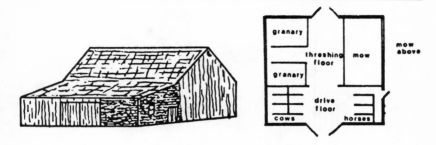

Fig. 6.34 The Acadian barn with its distinctive floor plan. (Courtesy of the American Review of Canadian Studies)

Range: Madawaska and St. John River valleys of New Brunswick and Maine.

CAJUN BARN

This frame barn has an 8–12 foot front opening centered on the gable end. (See fig. 2.4 for a typical Cajun barn.) The entry is recessed 6–8 feet into the barn. The side stables have a recessed corncrib in the center and a hayloft above. Sometimes there are attached side sheds. The barn is almost always 30-feet square. A small rear door is for loading corn. Older gable roofs of wooden shakes are now usually covered with tin. Monitor roofs replaced gable during the early 1900s. They were, in turn, replaced by gambrel roofs in the 1930s.

Range: Cajun settlement areas of Louisiana, extreme east Texas.

Variant 1. A longer barn, up to 57-feet long and 30-feet wide, with a freestanding corncrib (not visible from exterior) and attached shed roofed area for implement storage across the rear of the barn.

Range: Lower Bayou Teche, Louisiana.

Variant 2. A larger barn averaging 48-feet square due to augmented needs of sugarcane growing. Always with gable roof.

Range: Lower Bayou Teche, Louisiana.

Variant 3. Low pitch to roof as there is no hay storage. Entry at gable end. Large, raised central corncrib, no side stables. Roofed porch across front of barn, sides sometimes enclosed. Side sheds always present and usually with open fronts.

Range: French settlement, Louisiana.

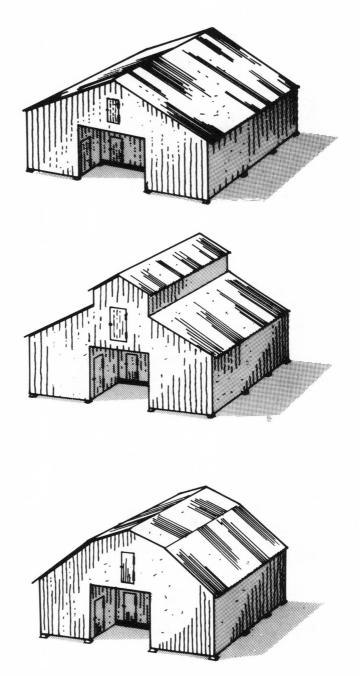

Fig. 6.35 Sketch of Cajun barns showing some of the possible roof variations. (Courtesy of M. Comeaux and the American Geographical Society)

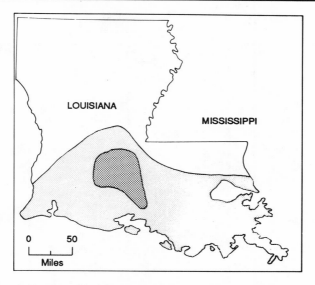

Fig. 6.36 Area where Cajun barns and other Cajun structures are most likely to be found. The darker area is the core of the region. (Courtesy of M. Comeaux and the American Geographical Society)

HISPANIC BARNS

In the Hispanic areas of northern New Mexico, three barn structures have been reported.

TASOLERA
This roughly built barn may house livestock, store hay, or perform both functions. The building itself is hard to describe because a wide variety of such structures exists. Most have gently sloping shed roofs. Associated with Hispanic settlement.
Other terms: Caballerisa, fuerte, tejavana.
Range: Northern New Mexico.

TAPEISTA
This simple structure consists of four posts and an elevated platform on which to pile hay or cornstalks. Later ones may have a crude open crib of horizontal log or vertical jacal construction under the platform. Built by Hispanic settlers.
Range: Northern New Mexico.

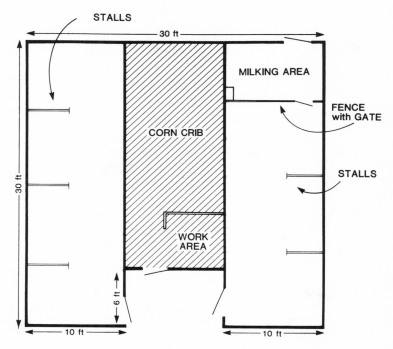

Fig. 6.37 The Cajun barn floor plan helps to identify these barns, which from the outside may look like several other types. The recessed entryway is the most distinctive feature. (Courtesy of M. Comeaux and the American Geographical Society)

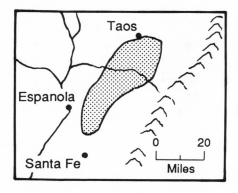

Fig. 6.38 The area of northern New Mexico in which Hispanic barns are likely to be encountered.

Fig. 6.39 The Tasolera is a primitive barn of Hispanic connection.

Similar structures: Mormon Thatched-Roof cowsheds, hay barracks.

HISPANIC TWIN-CRIB BARN
Two gable-roofed cribs of similar size and construction are separated by a roofed-over central breezeway or alley. The door placement varies: sometimes it is on the gable end, sometimes on the side end. Cribs are of horizontal logs. The earliest forms had single cribs, with a second crib added eventually; later the barns were built as complete structures. Similar in form to the Double-Crib barns of southeastern United States, but not related.
Range: Northern New Mexico.
Similar structures: Double-Crib barns.

OTHER ETHNIC BARNS

A number of other barns are associated with specific ethnic groups. The following list is by no means exhaustive; most of the research remains to be done.

DUTCH BARN
This compact, gable-front barn is squarish or often somewhat wider than long. It has large wagon doors, one or both of which are Dutch doors, and single, small doors near one or

Fig. 6.40 The Dutch barn.

both gable-end corners. The moderate-to-steep roof pitch means that the height of the ridge is more than twice the height of the low side walls. There is little or no projection of roof beyond the wall. Often a narrow pentice occurs over the wagon doors. Horizontal siding is usual, and sometimes a diamond-shaped window is placed high in the gable.

Range: Hudson, Mohawk, and Schoharie River basins of New York; Bergen, Somerset, Hunterdon, and Monmouth counties of New Jersey; southeastern Vermont.

Southwest Michigan Dutch Barn

This barn is larger than the eastern Dutch barn and with a rectangular plan. Often it has an asymmetrical (saltbox) roof, lower on the north or west. Frequently the barn is left unpainted, but with red-painted doors. Sometimes diamond windows are cut in the gable wall. There is vertical siding and often board-and-batten construction; the battens are painted white and the boards painted red.

Range: Allegan and Ottawa counties of southwestern Michigan.

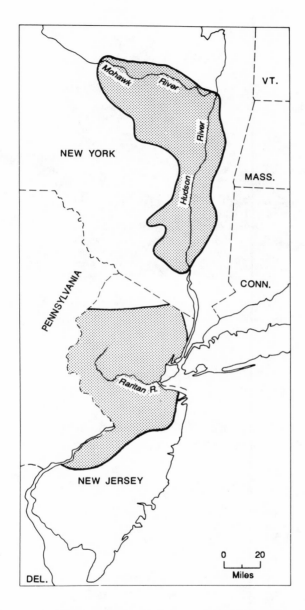

Fig. 6.41 Areas of New York and New Jersey where Dutch barns and other Dutch structures were located originally. Because of extensive suburbanization, only limited agricultural land remains.

Fig. 6.42 Southwest Michigan Dutch barn located near New Holland, Michigan. The squarish plan, wide roof expanses, diamond windows, gable doors, and vertical, white-painted battens are typical features.

SWEDISH BARN

An elongated, rectangular barn comprised of two disparate parts—a cow barn and a hay storage facility—it is typically 27–28 feet wide and 60–62 feet long. The cow barn section is

Fig. 6.43 The Swedish barn.

Fig. 6.44 *The Finnish Cattle barn. (Courtesy of M. Margaret Geib and the University of Massachusetts Press)*

usually constructed of tightly fitted hewn logs or fieldstone. The hay storage part of the barn is of timber frame covered by vertical boards or planks. The gable peaks are also covered by vertical boards.

Range: Chisago County, Minnesota; central Wisconsin; around Green Bay, Wisconsin.

FINNISH BARNS

Found throughout the upper Great Lakes area of northern Wisconsin, northern Minnesota, the upper peninsula of Michigan and western Ontario, they are of two distinct types.

FINNISH CATTLE BARN

This barn has a small, but relatively tall and narrow, gambrel roof, with a main door on the gable end. The siding and roof are vertical planks.

Range: Upper Great Lakes, Canadian Shield.

FINNISH HAY BARN

This small structure is made of loosely fitted, rounded logs held by saddle notches. Sometimes it shares a common roof with another structure, much like a Double-Crib barn. The most distinctive feature is that the side walls slope inward at the bottom to provide maximum rain protection.

Fig. 6.45 The Finnish Hay barn. (Courtesy of M. Margaret Geib and the University of Massachusetts Press)

Other term: Lato.
Range: Upper Great Lakes, Canadian Shield.

CZECH BARN
The dimensions are 27–30 feet by 48–80 feet, with an elongated rectangular plan, wagon doors on the gable end, one or two smaller doors on the side, fieldstone walls, and a gable roof. Small windows are usually high on the side walls. Sometimes the exterior walls are plastered and whitewashed.
Range: Southeastern South Dakota and perhaps other Great Plains states, but always in areas of Czech settlement.
Similar barn: Transverse Frame barn, but these are never built in stone.

Fig. 6.46 The Czech barn. (Courtesy of M. Margaret Geib and the Johns Hopkins University Press)

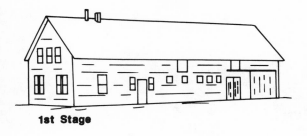

1st Stage

2nd Stage

3rd Stage

4th Stage

Fig. 6.47 Stages in the evolution of the Manitoba Mennonite housebarn. Originally built as a single unit, the house and barn were independently framed in the second stage. Both parts were further separated in the third stage, and the house was turned 90 degrees in the final stage.

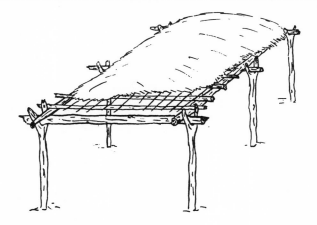

Fig. 6.48 The Mormon Thatched-Roof cowshed. (Courtesy of the California Folklore Society)

Variant. A very elongated barn, 44–160 feet long, with entrances on the side; it may be a modification of an Old World housebarn design. Even more rare than the basic barn.

OTHER REGIONAL BARNS

MANITOBA MENNONITE HOUSEBARN
The house and barn are separately framed. The house has solid walls of stacked horizontal two-by-fours or two-by-sixes. The barn is framed much like other timber-framed structures. Note the shed addition projecting sideways from the end of the barn. The ridgeline of the barn is almost always higher than that of the house.
Range: Southeastern Manitoba (southwest of Winnipeg); scattered locations in Saskatchewan and Alberta.

MORMON THATCHED-ROOF COWSHED
This simple structure consists of four to eight posts, connecting beams, and a thatched roof overall. The roof may be flat, or of a shed or very low-pitched gable type. The roof material consists of 1–4 feet of threshed loose straw. Sheds are often built along a board fence which provides shelter on one side.

Other walls formed of logs, planks, or railroad ties may or may not be present.

Range: Over the intermontane area between the Rocky Mountains and the Sierra Nevada, largely in areas originally colonized by Mormons.

Similiar structures: Tapeistas, hay barracks.

7

More Recent Barns

Toward the end of the nineteenth century, as farming became more scientific and less traditional, more efficient barn designs appeared. Because many of the designs were produced by state agricultural experiment stations and promoted widely by the agricultural press, these barns spread over vast areas and were adopted quickly by some farmers. But although these barns are widespread, their distribution is not as dense as that of ethnic barns. Because their designs are often more efficient than those of traditional barns, and because they were built later, these barns have been preserved in greater numbers than the ethnic barns. Gradually they may come to dominate the landscape.

THREE-END BARN
This barn evolved from English and Raised barns by adding a straw shed at right angles to the main structure. It may have a gable or gambrel roof, or a combination of the two. An extension may be up to one-half the size of the main barn structure. Eventually this barn was built as a complete unit beginning in the late nineteenth century. Both one-and-a-half-story and two-and-a-half-story versions exist. In Wisconsin this barn occurs primarily in an area of extensive Bohemian settlement. Other term: Three-Gable barn.
Range: Northeastern United States, eastern Wisconsin.

FEEDER BARN
At the end of the nineteenth century, farmers began to build barns using new construction materials and methods. Large timbers suitable for use in timber frame structures were in-

Fig. 7.1 A Three-End barn in Medina County, Ohio.

creasingly difficult to procure as agricultural settlement spread into the largely treeless prairies and plains of the Midwest. The perfection of the circular saw, the development of nail mills producing cheap nails, and the extension of railroads combined to make dimension lumber an economical construction alternative to pegged timber frames. A series of lumber trusses also permitted lighter frames. The barns thus evolved are usually referred to in the Midwest as Feeder barns. They have a rectangular plan, gable doors, and low-pitched roof.

Range: Throughout the Midwest and western United States; scattered in other regions.

Similar barns: Midwest Three-Portal and Transverse Frame, both of which, however, are timber-framed structures.

ERIE SHORE BARN

This small (30 by 40 feet) barn, with gambrel roof, is one story plus a loft. It has a side-to-side runway at the end of the barn and an off-center, side wagon door. Usually a row of low windows runs along the barn sides. The barn may have originated in the eastern Midwest about 1875.

Fig. 7.2 A Feeder barn from Dallas County, Iowa. This barn differs from a Midwest Three-Portal barn in that it is constructed of sawn lumber. Also it often has just one door on the gable.

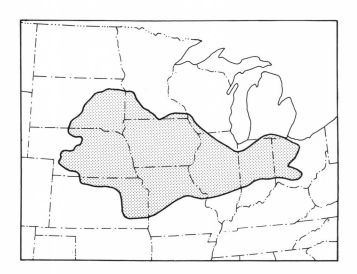

Fig. 7.3 Area within which Feeder barns are most often encountered.

Fig. 7.4 The Erie Shore barn. (Courtesy of M. Margaret Geib and the University of Massachusetts Press)

Range: Midwest and lower Great Lakes, southern Ontario. Similar barn: Feeder barn, which, however, has doors on the gable rather than off-center on the barn side.

ROUND-ROOF BARNS
Some consider this barn to be a type, but many different barn types have been roofed, or re-roofed, with round roofs formed of laminated rafters. Recent barn types, especially, are

Fig. 7.5 A Raised barn with a round roof.

Fig. 7.6 The Pole barn.

frequently found with round roofs, including the Erie Shore, Wisconsin Dairy, Foundation, and Three-End barns. Even Drive-in Crib, Raised, and English barns exist with round roofs. The roof became popular after World War I.

Other terms: Gothic, Gothic-Roof, Rainbow-Roof, Arched-Roof.

Range: Most common in areas of recent barns, especially in dairying areas where the increased hay storage of a round roof is advantageous.

POLE BARN

This one-story barn, with a concrete slab or dirt floor, has no loft. It is framed by upright poles inserted directly into the ground and characterized by wide gables and a low roof pitch. Later versions are of metal construction, especially those erected after World War II.

Other term: Loafing barns.

Range: Heavy concentration from Illinois to Ontario; widespread throughout northeast United States and eastern Canada.

Variant. One or two walls are left off so that cattle may have access to feed or shelter as they desire.

POLYGONAL AND ROUND BARNS

Promoted as efficient designs by popular writers of the mid and late nineteenth century, Round barns occur in both wooden frame and fieldstone. While an octagonal shape is most common, barns with various numbers of sides, from six to fourteen, have been identified. The barn may be one or two storied, with a ramp to the upper level. Usually used to house dairy cows.

Other term: Non-orthogonal plan barns.

Fig. 7.7 An Octagonal barn with a ventilating cupola. Such barns were typically built in the second half of the nineteenth century.

Range: Scattered throughout northern and central United States and southern Canada, especially in Ozaukee County, Wisconsin, Stephenson County, Illinois, and around Rochester, Indiana.

Variants. A handful of oval-plan and "doughnut"-plan barns exist in Douglas and Sibley counties, Minnesota, and Stephenson and Ogle counties, Illinois.

8

Special-Purpose Barns

Every student of buildings knows the saying "Form follows function." In other words, the purpose for which any building is built largely determines the shape of that structure. Simple barns perform only one or two functions. Multiple functions require a larger, more complex building. Because most barns shelter a few farm animals, store quantities of crops and grain, and keep equipment out of the weather, they have a basic resemblance to one another. Ethnic traditions are what separate and identify most of these barns. Some barns process a single agricultural product, and their form reflects this unusual role. Among the most conspicuous of these barns are those that shelter dairy cows or horses or that process or store potatoes, hops, and tobacco.

WISCONSIN DAIRY BARN

This large barn is 36 by 100 feet or larger. It has a gambrel roof or occasionally a round roof, although early versions were often gable-roofed with horizontal boarding. Note the rows of small windows and gable-end doors. There is usually a large gable-end loft opening and often a hanging gable (triangular hay hood). Frequently this barn has roof ventilators. Sometimes it may have a drive ramp to the loft level. Occasionally it is built as a bank barn.
Range: Minnesota eastward through Canada to New England and lower Great Lakes. In the so-called Dairy Belt.

MOUNTAIN HORSE BARN

This rectangular Western barn, usually about 28 by 40 feet, was described and mapped by J. T. Kilpinen in "The Moun-

Fig. 8.1 The Wisconsin Dairy barn. (Courtesy of M. Margaret Geib and the University of Massachusetts Press)

tain Horse Barn" (1994). Traditionally it is of round log construction, although hewn log and more recent framed examples exist. The main door is usually centered in the gable end, often with a large hay door above and a hay pole or hay hood. Typically there are two to five small windows in the

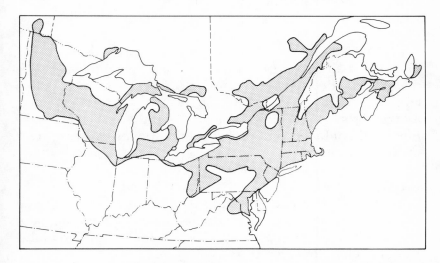

Fig. 8.2 Area within which Wisconsin Dairy barns are likely to be found. They are more common to the west of the region than to the east.

Fig. 8.3 A Mountain Horse barn from Montana. (Photo by Jon T. Kilpinen)

eave wall. A gable roof is most common, although recent examples often have gambrel roofs.
Range: Concentrated in northern Rocky Mountains and more widely in British Columbia.

POTATO BARN
This low, half-sunken barn usually has a gambrel roof. The design provides long-term, cool storage space for a potato crop.
Range: Aroostook County, Maine, central Wisconsin, Snake River Plain of Idaho, and probably other potato-growing areas of North America.

HOP BARN
Typically this two-part structure consists of a kiln or drying house and an attached processing and storage building (later versions incorporated both into a single building). Early kilns were circular stone structures with a high upper portion tapering to a vent protected by a distinctive cowl that could ro-

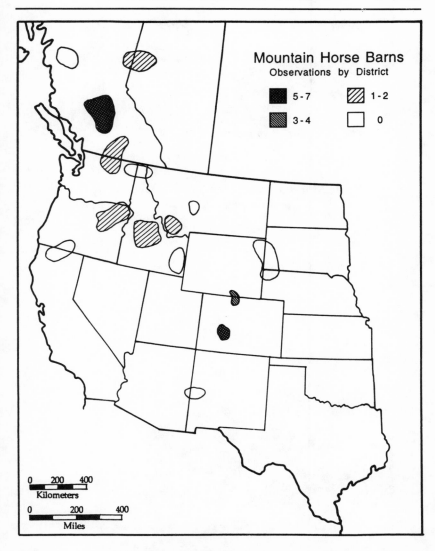

Fig. 8.4 Areas within which Mountain Horse barns can be found. (Photo by Jon T. Kilpinen)

tate according to wind direction. By the mid-nineteenth century the kiln was a square, two-story frame building surmounted by a cupola often on a steeply-pitched hipped roof. Range: Central New York; southern Wisconsin; northern California; Willamette valley, Oregon.

Fig. 8.5 Potato barn. Its low profile suggests its cool storage function. (Courtesy of M. Margaret Geib and the University of Massachusetts Press)

Fig. 8.6 Hop barns from central New York. The wooden, pyramid roof structure located in Madison County is more typical than the round, conical-roofed structure whose walls are made of cobblestone. This barn is located in southern Oneida County.

Fig. 8.7 The Fire-Cured Tobacco barn.

TOBACCO BARNS

Three major types are each associated with a different method of drying tobacco leaves.

FIRE-CURED TOBACCO BARN

Dimensions are often 20–22 feet by 26–48 feet, and 18–20 feet in height. Frame or log barns with gable entry.

Range: Western Kentucky and adjacent Tennessee, central Virginia.

FLUE-CURED TOBACCO BARN

Dimensions are usually 16–20 feet square, and over 20 feet high. This tall, squarish structure usually has one or more attached open sheds. Sometimes it is enlarged by constructing an adjacent square barn. The distinctive open sheds appear in an almost infinite variety.

Range: North Carolina, southern Virginia.

AIR-CURED TOBACCO BARN

This barn has much greater dispersion and variety than the other Tobacco barns. No special barn is needed for air-curing, so various barn types have been adapted to this use, especially Transverse Frame barns. The major adaptation has been the addition of frequent ventilator panels in the wall surfaces of the barn, usually vertical, but sometimes horizontal (especially in Pennsylvania). Horizontal ventilators are usually arranged so that several can be opened by one pole attached to the outside. Modifications to allow more roof ventilation are typical as well, including long horizontal openings, elongated

Fig. 8.8 The Flue-Cured Tobacco barn.

ridge ventilators, and clerestory roofs with side vents. Within a great diversity of adapted and purpose-built barns, several distinctive variants occur.

Range: Southern Ontario; Kentucky; Tennessee; south-central Wisconsin; Virginia; Darke County, Ohio.

Variant 1. The southern Ontario Tobacco barns usually occur in clusters of six to twelve identical squarish buildings.

Variant 2. The Tobacco barns of the Connecticut valley are extremely long and narrow. They occur singly but most often in pairs.

CARRIAGE HOUSE

This structure is really an urban barn. It is not often found on farmsteads where the barn performed its functions, which were to house carriages and provide stabling for a horse or two. They are most often seen today in smaller towns. Look for what appear to be tall garages.

Range: Widespread in the northern half of the United States and southern Canada.

OTHER BARN TYPES

Certainly as-yet-undescribed barn types exist. To qualify as a type requires that a barn form be found in more than a few

Fig. 8.9 Air-Cured Tobacco barns in southern Ontario near Tillsonburg.

Fig. 8.10 An Air-Cured Tobacco barn typical of south-central Wisconsin.
Note the metal ventilators along the ridge.

Fig. 8.11 A carriage house located in Allen County, Ohio. The Victorian trim is especially evident in this example.

Fig. 8.12 An Appalachian Meadow barn from near Clarksburg, West Virginia.

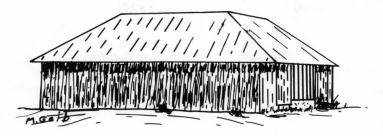

Fig. 8.13 The Wisconsin Hipped-Roof barn.

instances and that it deviate from known types in more than stylistic detail. One or two examples of a particular form may simply reflect the idiosyncratic style of a barn builder or farmer, especially if they are in close proximity to each other.

In West Virginia and southeastern Ohio, a small barn occurs widely which has not received any study. It seems to be something of a cross between the English barn and a small Transverse Frame barn. The small door is usually on the gable end. The building is rectangular, vertical sided, and unpainted. Its function varies from structure to structure, but includes stabling, equipment storage, and hay storage. Some of these barns which perform the latter function occupy isolated meadow locations.

North and west of Madison, Wisconsin, and extending over several hundred square miles, is found a type of one-story, square barn with a hipped or pyramidal roof. These barns look very much like a garage except that they range from 40 to 60 feet square and are vertically sided. They seem to be used for everything from housing dairy cattle to hay and machine storage.

Other Farm Structures

9

Secondary Farm Buildings

North American farmsteads normally contain structures other than the house and barn. Most of these secondary structures have agricultural functions, but some, such as woodsheds, privies, and summer kitchens, to name just a few, are more clearly related to the farmhouse and domestic functions than to the barn.

The arrangement of buildings on farmsteads varies widely, even within a small area. The careful observer, however, may detect certain basic arrangements. In much of western and central Canada and the United States, all buildings have the same alignment, to the cardinal compass points. This results, of course, from the rectangular land survey systems used in both of these areas. Even in eastern United States and Canada, early farm settlers often attempted to align their barns, houses, and secondary buildings, perhaps from an inbred sense of order.

Not all secondary buildings are so distinctly built as to be instantly or infallibly recognized. In these structures, the function has not determined the form of the building. A small sheep barn or a machine shed, for example, is hard to spot in the field. On the other hand, many buildings do betray their functions, and the motorist passing along the rural road can identify them.

CHICKEN HOUSES: A LESSON IN FIELD IDENTIFICATION

It isn't too much of an overstatement to say that virtually every farmstead in North America, at one time or another, had one or more structures to shelter chickens. It also isn't too far from the mark to say that most chicken houses cannot be identified by most observers. In many instances, these structures are small, nondescript buildings, often converted from an earlier, and quite different, use. The current function of these buildings may be hard to identify correctly in the field. There just aren't many distinctive external clues. Although their size is usually small, floor plan, roof type, walling material, and all other architectural aspects vary significantly from structure to structure.

Even when the chicken house has been built by farmers who have a strong ethnic orientation and who built them along recognized traditional building lines, few distinctive features are apparent. Much of the problem in being unable to comment on ethnic contributions to chicken house construction is that so few studies of the stuctures have been made. Only Malcolm Comeaux, investigating Cajun buildings in Louisiana (*The Cajun Barn*, 1989), and Amos Long, Jr., looking at German settlement in Pennsylvania ("Pennsylvania German Family Farm," 1972), have produced serious studies. Until further research is carried out, little that is definitive can be said about how the various ethnic groups approached the building of chicken houses. And if Cajun and Pennsylvania chicken houses are typical, few features easily recognizable in the field will be apparent.

In only one location in all of North America have chicken houses been examined to try to identify the entire range of such buildings. John Passarello, in an article for *California Geographer* (1964), noted that at least six distinct types of chicken houses exist in the Petaluma area of Sonoma County, just north of San Francisco. Each type was characteristic of a particular time period. No other area in all of North America has received similar attention, although disparate structures exist. In Wisconsin, chicken houses frequently have half-clerestory or half-monitor roofs which identify them.

In the twentieth century, as chicken farming, along with other kinds of agriculture, became more specialized, chicken houses became more standardized and recognizable. Elongated, low, gentle-pitched gable-roofed buildings primarily of pole construction, they dot the agricultural landscape from coast to coast, but not to the exclusion of other structures housing chickens.

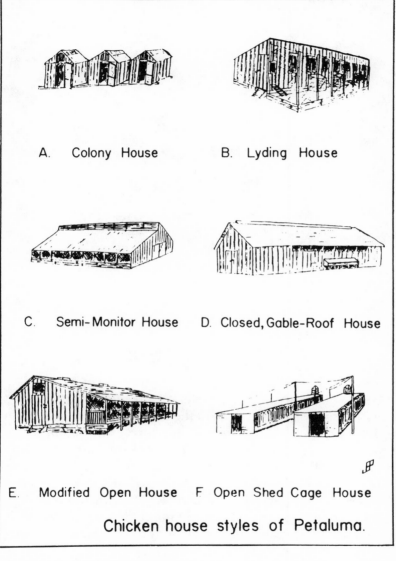

A. Colony House   B. Lyding House

C. Semi-Monitor House   D. Closed, Gable-Roof House

E. Modified Open House   F. Open Shed Cage House

Chicken house styles of Petaluma.

Fig. 9.1 Examples of chicken houses, Petaluma County, California. (Courtesy of the California Geographer*)*

Fig. 9.2 A Midwest storm cellar located near Livonia, Arkansas. The door is made of heavy corrugated iron. A thick cement slab covers the cellar and is pierced with ventilation pipes.

One of the lessons to be gained from all this is that we should be tentative in assigning functions to buildings. We will never be able to identify every farmstead structure, especially just by looking at it from the outside. Second, functions and uses change, so that clues may turn out to be false. Perhaps most important, not all farm structures have been studied sufficiently so that they can be readily identified.

What is true of chicken houses is also true, albeit usually to a lesser extent, for other farmstead buildings. Please keep these caveats in mind as you use the materials in this section to identify the wide range of secondary structures which appear on Canadian and American farms.

STORM CELLARS
Designed to offer protection from tornadoes and other severe storms, these structures were excavated into hill slopes or mounded up in level ground. The hatch-style doorway is usually oriented to face north or east.
Range: Midwest, Great Plains, Prairie provinces.

Fig. 9.3 The privy. The building is tall, squarish, and usually not confused with any other farmstead building.

PRIVIES

The small, but tall, wooden "outhouse" or toilet was separated from the house to protect against odor, but it was kept close enough for convenience in inclement weather. Usually it had a roof ventilator and decorative cutouts on door. The lower part of the back wall was sometimes hinged to allow for easier cleaning.

Other terms: Outhouse, back house, necessary (East Coast).

Range: Everywhere, but becoming harder to find.

10

Food and Water Storage

A number of buildings on North American farmsteads are associated with food or water storage. These structures are usually quite distinctive and hence are easily recognizable in the field, even though the particular form may vary somewhat.

SPRINGHOUSES

This small structure was designed to protect the source of spring water and to provide cool, clean storage for dairy and perishable farm products. Usually it is indistinguishable from other farm buildings except that the springhouse is generally located at the base of a slope and is often built into the slope. Louvers and small roof ventilators as well as external water outlets may serve to identify this building. Usually a springhouse has masonry construction, or at least the bottom part is stone.
Range: Widespread over entire continent.

MILK HOUSES

Milk was originally stored in springhouses, but government regulation and improving commercial standards forced an improvement in milk cooling methods. Specifically, milk must now be cooled to 50 degrees or lower within a few hours of milking to prevent bacterial growth. Rectangular in form and gable-roofed, the milk house is an unmistakable trademark of the dairy farm. Located as close as possible to the barn itself, it is often an appendage of the barn. By law, however, the milk house must be separated from the barn for sanitary reasons. The milk house is small, now usually made of concrete or tile blocks, and well insulated with little or no ventilation.
Range: Dairy belt of northeastern North America.

Fig. 10.1 A typical springhouse located at the bottom of a slope and well away from possible pollution from the barn and feedlot. Dauphin County, Pennsylvania.

FARM WINDMILLS
The American farm windmill dates from the mid-1800s, when it was used widely to lift well water to livestock and later to provide electricity. Only the blades and rudder rotate, allowing the windmill to work constantly, regardless of wind direction. These structures are often derelict today, doomed by rural electrification except on Amish farms.
Range: Widespread from its origins in Connecticut and California, throughout Canada and the United States. Especially important in Great Plains and Prairie provinces, and elsewhere as an indicator of Amish settlement.

DOMESTIC TANKHOUSES
At its simplest, this structure consists of a large wooden tank, elevated (30–40 feet) on an open, sturdy wooden frame to ensure gravity flow. Frequently it is boxed-in within a roofed enclosure. The enclosed frame provides for one or more rooms under the tank. A variety of subtypes occur, some of

Fig. 10.2 The milk house is usually adjacent or close to the barn. Its entrance must be direct from the outside, not through the barn. The barn is a good example of a Raised Gothic-Roof barn. Note also the gable-end pent. (Courtesy of M. Margaret Geib and the University of Massachusetts Press)

which are incorporated into the architecture of the farmhouse itself.

Range: Valleys of California, Willamette valley, Great Plains, eastern Texas, widely scattered over the Midwest.

CAJUN CISTERNS

These raised rainwater storage tanks are usually made of wood, less often of brick. Masonry or brick piers raise the tank and allow a fire to be built underneath. They are usually sited at the rear corner of the house to collect rainwater runoff from the roof. See figure 6.36 for most likely areas of occurrence.

Range: Cajun Louisiana.

CELLARS

Cellars provide storage for root crops as well as other vegetables and foodstuffs. Mostly they are excavated and below ground to provide maximum insulation. Frequently they can be identified by a sloping door (or doors) against a bank, and

Fig. 10.3 Domestic tankhouses and the areas in California where each kind is most common. (Courtesy of M. Margaret Geib and the University of Massachusetts Press)

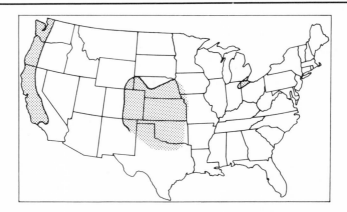

Fig. 10.4 Areas where domestic tankhouses are likely to be found. The eastern boundaries of the large midwestern area are not yet clearly defined.

often a ventilation pipe projects above the ground. Cellars also provide protection against strong winds and tornadoes, which occur especially in the Midwest. Usually they are constructed of stone or mortar. See figure 9.2 for the similar-appearing storm cellar.

Fig. 10.5 A Cajun cistern near Napoleonville, Louisiana. (Courtesy of M. Margaret Geib and the University of Massachusetts Press)

Other terms: Root cellar, storm cellar, cold cellar, canning cellar, or cyclone cellar.

Range: Widespread, most common in Great Plains and Midwest, extending into southern Ontario.

Variant. Ramps leading to Raised barns occasionally are hollowed out to accommodate the cellar.

11

Food Processing

The structures described in this chapter function to process food in some way rather than primarily to store it, although some overlap of functions is inevitable. These structures generally require more careful observation to reveal their identity.

SUMMER KITCHENS

Sometimes an earlier, cruder dwelling was converted to a summer kitchen when a larger farmhouse was built. More often, it is a separate, simple, rectangular building, closely situated to the main house kitchen and built specifically as a summer kitchen. Early versions with large fireplace and chimney were later replaced by ones with stove and pipe. Often an open cupola with a dinner bell is centered on the ridgeline; sometimes the bell tops a pole in the farmyard near the summer kitchen. Detached summer kitchens are associated with settlements of Pennsylvania Germans, Hungarians, French-Canadians, Belgians, Russian-Germans, and Finns. Most immigrants from the British Isles did not build summer kitchens.

DRYHOUSES

This small (up to 8 feet square) wooden frame outbuilding was used to dry and preserve vegetables and fruit. Shallow trays in the upper part of the house could be slid out and filled from the outside. The chimney flue is in the gable end. This structure, known from the early German farms in Pennsylvania, is rarely encountered today, although one example exists near Elletsville, Indiana.

SMOKEHOUSES

This structure is usually a small wooden-frame outbuilding with no windows and a small door in the gable end, with

Fig. 11.1 The summer kitchen is usually sited close to the rear door of the farmhouse for convenient access. Note on this example from central Ohio the pent roof over the gable-end door.

small flue openings under the eaves or in the gable. The wall material can be log, hewn timber, cobblestone, brick, stone, or stackwood. Several types of smokehouses probably exist, although additional research is sorely needed and quickly before remaining structures disappear.

ENGLISH TIDEWATER SMOKEHOUSE

Only brick and stone examples are extant, although originally frame construction probably was more common. Square or rectangular plans, 7–11 feet on a side.

Range: Tidewater Virginia, Maryland, New Jersey.

DUTCH SMOKEHOUSE

This structure is square, between 6 feet and 8 feet, with a large chimney in the rear gable wall. Usually of brick construction.

Range: New Jersey, Hudson valley of New York.

PENNSYLVANIA GERMAN SMOKEHOUSE

This structure is square or rectangular, 6–8 feet on a side, with a height of 8–12 feet. It is constructed of frame with board and batten. Some are stone and brick, but these are widely scattered.

Fig. 11.2 Examples of dry-houses. (Courtesy of M. Margaret Geib and the University of Massachusetts Press)

Range: Eastern Pennsylvania; scattered elsewhere in places of German settlement.

UPLAND SOUTH SMOKEHOUSE

This structure is an outgrowth of English and German smokehouse designs, but with a strong Scotch-Irish influence. The early ones are of log, the later ones of box or frame construction. Typically one measures 6–10 feet per wall and may be square or rectangular in plan. Sometimes one is built over a root cellar. Look for whitish salt residue on lower walls.

Range: Central and southern Appalachia.

CANTILEVERED-ROOF SMOKEHOUSE

Derived from the Pennsylvania German smokehouse, this structure is larger, at 12–14 feet on a side and rectangular in plan. Its cantilevered roof extends over the front gable to shel-

Fig. 11.3 A smokehouse in Perquimans County, North Carolina. Smoke-houses are often difficult to identify, although their squarish floor plan and lack of windows are good clues.

ter the door. The height is 8 feet to the eaves. Early versions were of log construction.

Range: Scattered through Appalachia from Pennsylvania to East Texas.

TWO-STORY SMOKEHOUSE

Derived from the English Tidewater smokehouse, this structure has two stories, with a height of 16 feet to the eaves and often a pyramidal roof. Constructed in lumber, brick, and stone, less often of logs.

Range: Scattered through Appalachia from Virginia to East Texas.

BAKEOVENS

In areas of English, Scotch-Irish, and Dutch colonial influence, bakeovens were generally built into the kitchen hearth inside the house. However, built-in bakeovens beside the fireplace are virtually unknown in the Upland South region,

Fig. 11.4 A bakeoven with a partially enclosing shed over it, most typical of Pennsylvania bakeovens. Elsewhere the protective enclosure is usually lacking.

perhaps because of the warm temperatures of that region. The French, Belgian, Spanish, German, and other settlers built separate bakeoven structures.

SPANISH-MEXICAN BAKEOVEN

Adobe or stone covered in adobe, this oven is beehive-shaped. Rarely is it larger than 4 feet in diameter, and it is raised on a one-foot-high plinth of stone or adobe. Often it occurs in groupings. Also used to finish pottery. Adopted by Pueblo Indians.

Other terms: Horno.

Range: Southwestern United States.

CANADIAN FRENCH BAKEOVEN

Brick or stone is more commonly used to build this oven than clay. If clay is used, usually a wooden shed or gable roof covers the oven to protect the erosion-prone, domed clay structure. Ovens are oval, pear-shaped, or rectangular and raised on wooden platforms.

Range: Quebec, New Brunswick, Nova Scotia.

Variant. This rare, massive stone bakeoven has a large chimney at the front. The vault is usually of brick and covered with a wooden hipped or gable roof resting on top of stone walls surrounding the vault.

BELGIAN BAKEOVEN

Made of brick and/or stone, this oven is usually attached to the rear gable of the summer kitchen and accessed from the inside of the kitchen. The bakeoven is set on platform of limestone, 6–7 feet square and about 4 feet high.

Range: Door and Kewaunee counties, Wisconsin.

GERMAN BAKEOVEN

Usually this oven is a separate, free-standing structure, although sometimes it is attached to a summer kitchen and/or to the main house fireplace. Where the bakeoven is attached to a structure, it extends beyond the wall and often is covered with a wooden roof. A large structure of 8–12 feet, this oven includes a 3-foot roof overhang toward the front. Normally only the triangular projecting overhang is enclosed, but sometimes sidewalls and even an end wall with door enclose the projection. The gable roof is wooden shingle or slate and is supported by corner posts elevating the roof 3–4 feet above the oven. The brick chimney is at the rear.

Range: Pennsylvania.

Variant 1. Chimney and flue located at the front of the bakeoven.

Range: Eastern Pennsylvania.

Variant 2. Chimney at front of the bakeoven, but flue opening is at the rear and connected by a curved flue. Called a squirrel-tail oven.

Range: Bucks County, Pennsylvania.

SORGHUM MILLS

The sorghum mill consists of two structures—the crushing mill and the evaporator—which work together in the production of sorghum syrup. The structures are usually found in close proximity. Sometimes a summer kitchen or a main house fireplace is used instead of an evaporator.

Range: Appalachia, Ozarks, southern Wisconsin.

CRUSHING MILL

A set of wooden or metal rollers supported on wooden stumps or a timber framework, 3–4 feet high. Gears and vertical shaft are attached to a long horizontal boom or sweep powered by horse or mule.

Fig. 11.5 A sorghum mill on the grounds of the Old World Wisconsin Outdoor Museum.

Fig. 11.6 A sugarhouse in Geauga County, Ohio. Steam pours out of the elongated ridge ventilator, as well as the door, as the maple sap is boiled down. Note the collecting pails attached to the maple trees.

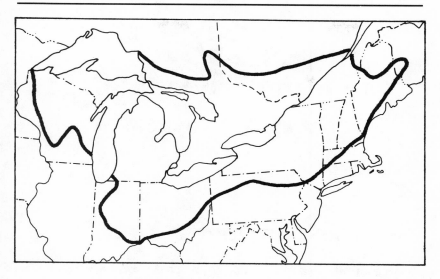

Fig. 11.7 Areas where sugarhouses are likely to be found. They are typically located in groves of maple trees and not with other farm buildings.

EVAPORATOR

A deep brick or stone fireplace, 4 feet wide by 12 feet long, a metal chimney 8 feet tall, and a slightly tilted evaporating pan resting on the fireplace.

SUGARHOUSES

A sugarhouse is usually located in the sugar bush, a grove of maple trees, and not in the farmstead itself. Houses range from small, one-room structures of 10 by 18 feet, up to two-room houses of 16 by 36 feet. A chimney or stovepipe is at one end. A distinctive feature is the large elevated roof ventilator, with open louvers or hinged doors. The ideal location for a sugarhouse is at the foot of a small slope, thus allowing gravity to feed sap into the evaporator. Cordwood stacked along one wall is another characteristic.

Range: From northeastern North America westward into the upper Great Lakes; especially common in northern New England, eastern Ontario, Quebec, New Brunswick, New York, northeastern Ohio.

12

Grain and Fodder Storage and Hay Processing

GRAIN AND FODDER STORAGE

A large number of buildings, or structures that are not buildings, are devoted to processing such field crops as small grains, corn, or hay. In eastern North America, silos are so common in dairying areas that many people erroneously associate silos with all farms. Also silos are a relatively recent addition to the farm landscape, dating from around the turn of the twentieth century. Corncribs, on the other hand, are much earlier, although today only the huge metal corn storage bins are still commonly seen.

In western North America, probably because of the drier climate and hence the better environment for wood preservation, some early devices for hay processing survive in reasonable numbers. They are often encountered in the more isolated mountain valleys.

GRANARIES

These small, rectangular, gable-roofed structures are used for storing small grains such as wheat, barley, and oats. A distinctive feature is the lack of windows or other openings in order to make it difficult for vermin to enter. Often they are double-walled for maximum grain protection. The most distinctive feature is the elevation of the building on several short piers of wood, stone, or cement block. Sometimes metal disks are added to the piers under the sills of the building to make the granary animal-proof. Granaries seem to be associated with settlement of German, Scandinavian, and eastern European migrants.
Range: Eastern and central North America.

Fig. 12.1 *The granary is normally elevated on posts or piers and has as few openings as possible. This one is in Lorain County, Ohio.*

CORNCRIBS

A wide variety of designs and materials are used in both home-made and manufactured corncribs. All designs must allow the moist, newly harvested ears of corn to dry slowly and steadily in order to reduce losses from mold and mildew. The walls must contain a high proportion of open area, usually attained by use of widely spaced, narrow slats. The structure itself must be narrow in order to ensure adequate air circulation, or the corn must be artificially dried. The narrower the crib, the better the drying process. The proper width of an ordinary crib in a particular locality depends on the date at which corn normally matures and on the prevailing weather conditions during the first eight months of storage. Early versions were built of small-diameter, unhewn, or split logs, later versions of narrow lumber slats. The crib usually rested on log or stone piers. The types listed are preliminary as there have been few studies of corncribs.

GABLE-ROOFED CORNCRIB

This most common type of corncrib has vertical or horizontal slats. The gable roof usually has loading doors and sometimes ventilators.

Range: Eastern United States and Canada.

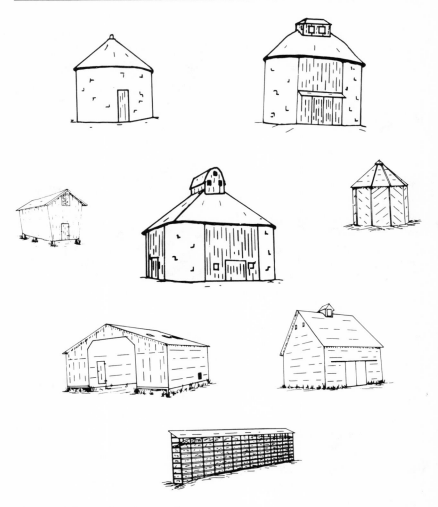

Fig. 12.2 Examples of corncribs.

SLANT-SIDED CORNCRIB

The gable roof with sides slanting inward to the base affords weather protection and assists with unloading. This type was very common throughout the eastern United States and Ontario in the late 1800s. Its small size makes it impractical today.

Other term: Connecticut corn house.

Range: Eastern United States, Midwest, New England, Ontario.

SHED-ROOF CORNCRIB
This structure is long and narrow, with a shed roof and vertical sides. Wire mesh is often substituted for slats. Frequently it is oriented north-south to catch the prevailing winds and drying sunlight. Developed in the late nineteenth century.
Range: Midwest, eastern United States, Ontario.

DRIVE-IN CORNCRIB
This more elaborate version of the corncrib is actually a barn, the Drive-in Crib barn (see chap. 4).
Range: Midwest, eastern United States, Ontario.

CIRCULAR CORNCRIB
This structure is almost always made of wire mesh with a conical, metal roof. The openness of the wire sides permits the thicker diameter of the round form without loss of drying capability. Developed in the early twentieth century.
Range: Midwest, eastern United States, Ontario.

MIDWEST MASONRY CORNCRIB
These gigantic masonry or clay tile structures were built from the 1920s through the 1950s. The blocks are pierced with rows of narrow slits. Often as large and high as a house, these structures range from simple circular cribs through double cribs to an enclosed structure of four internal cribs.
Range: Iowa, Illinois, and elsewhere in the Corn Belt, but less often.

SCIOTO COUNTY ELEVATED CORNCRIB
This large and much elongated (up to 120 feet) drive-through crib is raised up on stone pillars, with a gable roof and partially open, diagonally slatted sides. Lumber frame, louvered openings for unloading.
Range: Apparently found only in the middle Scioto valley in Ohio.

SCALE HOUSE
This simple, tunnel-like structure with open gables covers a scale for weighing corn. A shed-roofed addition houses the room where the scales were read and records were kept.
Range: Scioto valley, Ohio; scattered elsewhere throughout the Corn Belt.

SILOS
Covered pit silos were first widely built in North America in the early 1880s. The earliest upright or tower silos date from

Fig. 12.3 A Scioto valley corncrib, elevated on concrete pillars. Note roof loading-hatches and gable wagon-entry.

the end of the 1880s and round or circular forms from the late 1890s. The shift from the rectangular to the circular form stems from the efficiency of the circular form in storing corn ensilage by eliminating air space and thereby reducing spoilage. The first silos were pits excavated *inside* the barn. Later, silos were usually built adjacent to the barn on the gable end. Early roofs, even on circular silos, were gable roofs. Roof forms better suited to the circular silo evolved through conical, hipped-conical, low dome, to hemispherical. The following silo types reflect primarily the construction material and make up a roughly evolutionary sequence.

RECTANGULAR WOODEN SILO

In this earliest upright form are found the same materials and techniques as those used in the barn itself. The silo was constructed in some areas as late as 1910 and has framed lumber walls. Later examples frequently have corners rounded off on the inside with a vertical tongue-in-groove lining.
Range: Northeastern Ohio, southwestern New York, Puget Sound lowland, Wilamette valley of Oregon; scattered elsewhere.

OCTAGONAL SILO

This uncommon form attempts to achieve the advantages of a circular silo while keeping the ease of angular construction.
Range: Rare but widespread; small concentration in Vermont.

Fig. 12.4 A scale house from Pike County, Ohio. The shed to the left contains the weighing mechanism.

WOODEN-HOOP SILO
The wood of this silo was soaked and shaped into gigantic circular hoop forms and then fastened together horizontally in the tower shape. This style did not become popular because the hoops tended to spring apart.
Range: Rare, but widespread. More likely to be found in Wisconsin than elsewhere.

WOODEN-STAVE SILO
The tongue-in-groove, vertical wooden staves of this silo are held in place by iron bands and turnbuckles. Built from 1894 onward.
Range: Common throughout the Dairy Belt.

FIELDSTONE SILO
This early variant was expensive to construct and limited to those areas of abundant fieldstone.
Range: Mostly in glaciated areas of northeastern United States.

MASONRY SILO
Usually concrete blocks, but sometimes more expensive tile blocks, were used to build this silo. Brick construction was

Fig. 12.5 Examples of silos.

very rare. Frequently it was topped by a distinctive, low-domed masonry roof. Dates from around World War I.

Range: Throughout Dairy Belt, but especially in Wisconsin.

POURED CONCRETE SILO

This silo was formed of separately poured, stacked, concrete rings an built in the early twentieth century. Like most earlier silo types, unloading was from the top. A ladder of metal rings sometimes enclosed by a wooden projection is often found on the outside.

Range: Entire Dairy Belt.

CEMENT-STAVE SILO

Similar in construction to the Wooden-Stave silo, but with cement staves, this structure was perfected by cement companies about 1906.

Range: Widely diffused through the United States.

HARVESTORE SILO

Invented by the A. O. Smith Company in Milwaukee, Wisconsin, after World War II, this silo is assembled with fiberglass and metal panels. Other companies have produced similar metal silos. All these silos empty automatically from the bottom. Larger than other silo types.

Range: Familiar blue silos nearly ubiquitous in eastern North America.

HORIZONTAL SILO

The trench (below ground) or bunker (above ground) silo dates from the end of World War II. Geared to grass ensilage, mechanical harvesting, and self-feeding, it is often sealed by sheets of plastic held in place by large numbers of used tires.

Range: common only on the Great Plains, but scattered throughout the Dairy Belt.

HAY PROCESSING

HAY BARRACKS

Used to store hay and straw, this simple structure has four corner posts and a pyramidal or gable roof, with the gable ends enclosed. The foundation consists of light sills laid on the ground or placed on cornerstones. Pegs in the corner post allow the roof to be raised or lowered with a ratcheting jack.

Fig. 12.6 The hay barrack.

This structure has strong associations with Dutch settlements as well as some German and Ukrainian.

Other term: Shotscheier, among the Pennsylvania Germans.

Range: Although now very rare, they have been reported in scattered locations in eastern Massachusetts, Virginia, Maryland and Ohio, eastern Iowa, northern Illinois, western Wisconsin, western New York, Rhode Island, Prince Edward Island, southeastern Manitoba, Irish farms in eastern Newfoundland, and especially Dutch farms in the Hudson valley and northern and central New Jersey.

HAY DERRICKS

Associated so completely with Mormons, the hay derrick has been used as an index to identify Mormon settlement. The structure—a pyramidal base with attached boom—uses ropes and pulleys, with horsepower, to stack hay so as to reduce spoilage. Seventeen types of Mormon hay derricks have been identified and are shown in figure 12.7, along with locations where they have been reported.

Range: Utah and Mormon settlement areas in surrounding states.

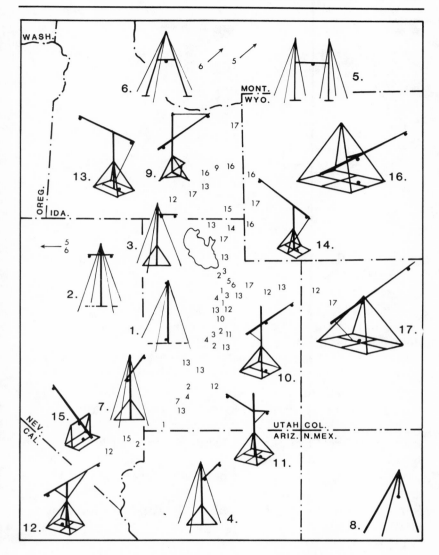

Fig. 12.7 Examples of hay derricks. The smaller numbers indicate places where the different types can be found. (Courtesy of M. Margaret Geib and the University of Massachusetts Press)

HAY STACKERS

This structure, used to stack hay, is of quite different design than a hay derrick. Gradually it was replaced by modern hay balers after World War II. Several subtypes can be identified.

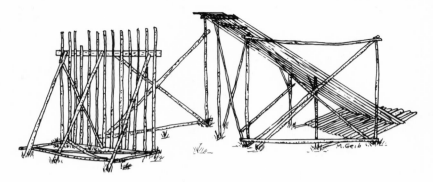

Fig. 12.8 A Beaverslide hay stacker.

Range: Western North America, especially in Montana, Wyoming, and Alberta.

RAM STACKER

This fan-shaped slide was made of long, straight poles of small dimension held aloft at a 45-degree angle and a racklike device in which the hay is held.

Range: Throughout the northern Great Plains and southern Prairie provinces.

BEAVERSLIDE STACKER

The name is derived from the Beaverhead County Slide stacker, which refers to its place of origin in the Big Hole Valley of western Montana. This slide is made of long, straight poles or narrow boards within a frame of heavier poles, with a 45-degree angle and a two-sided hay basket pulled by cable and pulley up the slide.

Range: Especially in Montana, Wyoming, Colorado, and Utah.

OVERSHOT STACKER

This stacker functions like a catapult. The hay basket is mounted on two long poles, the ends of which pivot on the base. The base is a rectangular skid arrangement with a pyramidal superstructure. With cables and pulleys, the hay basket is raised in an arc, up and over the base.

Range: Great Plains and Prairie provinces.

SWINGING STACKER

This stacker is similar in form to the hay derrick. Its pyramidal base of skids and poles has a long counterbalanced pole attached. The long end of the pole supports the hay basket,

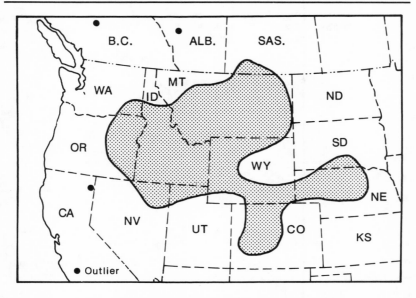

Fig. 12.9 Area within which Beaverslide hay stackers can be found.

while the short end supports a wooden box filled with boulders. The hay load is swung sideways and elevated, and the hay can be deposited at any part of the stack.

Range: Great Plains and Prairie provinces.

COMBINATION STACKER

Combining the functions of the sweep rake and stacker in a single unit, this stacker resembles a crude front-end dirt mover. With the hay basket in front, the base and pyramidal frame are wheel-mounted and can be moved. An elevating mechanism is engaged as the stacker approaches the stack until it is high enough for the hay to be deposited.

Range: Great Plains and Prairie provinces.

13

Fencing

Fences are perhaps more widely distributed than any other rural cultural element of the landscape. Fence types and their evolution in North America are closely tied to the physical environment, to availability of materials, to changes in land use, and to technological innovation.

BRUSH FENCE

Made of piles of dead branches and trees removed during land clearing, this fence takes up a great amount of land. It is impermanent and rarely encountered today.

Range: Northern Michigan and elsewhere on marginal agricultural land.

STUMP FENCE

This fence, constructed of uprooted tree stumps, generally with the roots facing outward, is similar to the brush fence and also usually related to land clearing. It is more permanent than the brush fence but still not common. Usually this fence is not found in poorer agricultural areas because the return from the land does not warrant the investment.

Range: Widely scattered, but rarely encountered.

STONE FENCE

Often related to land clearing, especially in glaciated areas, this fence is more boundary marker than fence due to the difficulty of piling stones to any great height. Usually made up of irregular, rounded fieldstones.

Other term: Rock fence is the name used throughout the Upland South region.

Fig. 13.1 A brush fence from Montcalm County, Michigan.

Range: Areas of glacial deposition in northeastern North America and the Great Lakes.

Similar feature: The "rest" stones left behind after a rail fence (see below) has vanished might be mistaken for the remains of a stone fence.

STONE WALL

Not to be confused with the stone fence, the stone wall is more nearly vertical and carefully laid. At least four subtypes exist, and they depend in part on available stone in an area. The term *stone wall* is used in the North, whereas the term *rock wall* is used in the South for the same feature.

MORTARED STONE WALL

Fieldstone laid up in mortar. Made up of irregular boulders and stones.

Range: Occasional in areas of glacial deposition.

STACKED STONE WALL

Restricted to areas of flat stone, especially limestone (sometimes sandstone, slate, or shale). Flat stones are laid up hori-

Fig. 13.2 This example of a stump fence is erected at the Old World Wisconsin Outdoor Museum in Eagle, Wisconsin.

Fig. 13.3 A typical stone fence from Middlesex County, Massachusetts. Note the care with which the rocks have been placed.

Fig. 13.4 A stone or rock wall from Bourbon County, Kentucky. Note how closely the rock pieces fit together and the essentially vertical orientation of the topmost course of rocks.

zontally. Both dry stone and mortared types exist. Sometimes stones are laid in two side-by-side rows with rubble or concrete filler. The top row of stones may be set on edge vertically.

Other terms: Rock fence, stone row, plantation fence, turnpike fence.

Range: Limestone areas, including Bluegrass Basin of Kentucky, southern and eastern Wisconsin, and Texas Hill Country.

SLAB WALL

Slabs of sandstone built of two straight faces, with the irregular space between filled with rock fragments and rubble and capped with large, flat, overhanging stones.

Range: Northeastern Pennsylvania.

EDGE FENCE

Flat limestone blocks stacked diagonally and without mortar. Not very common, found usually in hillier terrain.

Other term: Rick-rack.

Range: Kentucky Bluegrass basin.

Fig. 13.5 The log-and-chock fence is still common in the valleys of the Rocky Mountains. Long logs alternate with short sections. (Courtesy of M. Margaret Geib and the University of Massachusetts Press)

LOG-AND-CHOCK FENCE

For this early pioneer fence, found in wooded areas, cleared trees were used with little trimming. Stability was provided by cutting some logs into two- to three-foot chocks, which were inserted, more or less at right angles, between the logs. Sometimes incipient saddle notching was used to add stability. Possibly of Swedish origin.

Range: Now mostly found in mountainous western Canada and United States, but also common in northeastern New Brunswick.

RAIL FENCE

In this fence most closely identified with eastern pioneering areas, the courses of split rails (sometimes saplings) intersect the next section at about a 120-degree angle. The lowest rails are often supported by fieldstones. These "rest" stones sometimes remain in place after the wooden part of the fence has deteriorated. The rail fence was proclaimed the "national fence" by the U.S. Department of Agriculture in recognition of its widespread use in the late 1800s.

Other terms: Virginia, split-rail, worm, snake or zigzag fence.
Range: No other fence was so widely distributed throughout the entire United States and Canada in the nineteenth century before the introduction of wire fencing. Still common in cen-

Fig. 13.6 Examples of rail fences. The earliest type is to the left. An improved type using uprights for additional support is to the right.

tral Ontario, central Pennsylvania, western Virginia, and generally throughout Appalachia.

Variant. Vertical supports added at the intersection of the rails and sometimes wired together clasping the rails.

STAKE-AND-RIDER FENCE

With rails supported by a pair of crossed stakes, this fence requires no postholes, but the stakes must be bound together. One sturdy rail, the rider, is laid on top of the fence in the crotches of the stakes to provide stability at each crossing. Sometimes a rock is placed in the crotch atop the rider to help anchor it. Possibly of Finnish origin.
Range: Southern part of Canadian Shield, central Appalachia, north-central Pennsylvania.
Similar fence: Russell fence (see below).

RUSSELL FENCE

The montane West has limited timber, and trees are small and twisted. Soil is often thin or nonexistent, so postholes are difficult to dig. The patented Russell fence solved those problems. It is clearly related to the stake-and-rider fence above. Four posts, two at a high angle and two at a more oblique angle, are secured

Fig. 13.7 A stake-and-rider fence near Branford, Ontario.

by strong wire. The top rail is secured in the crossing of the two sets of posts. All other rails are suspended by wire secured from the post junction or from a higher rail.

Range: Interior areas of Pacific northwest, including British Columbia.

IRISH FENCE
Somewhat similar to the stake-and-rider, the Irish fence involves crossed stakes, but only one end of a single pole rests on the crotch, which is then wired together, while the other end rests on the ground.

Other terms: Shanghai fence, Swede fence, buck fence, reindeer fence.

Range: Northern Arizona, southern Appalachia.

SAWYER FENCE
Similar to the Irish fence, the sawyer fence uses slabs of wood left over from sawmill operations (hence the name) in place of poles. Given that the slabs are much shorter than poles, the angle of the cross-members of the fence is much greater than in the Irish fence.

Range: Very scattered and rarely encountered.

Fig. 13.8 Section of an Irish fence near Payson, Arizona.

POST-AND-RAIL FENCE
Two or three split rails are mortised into upright timber posts. Other term: Called a pieux fence in the Cajun country of Louisiana.
Range: In 1871, according to federal statistics, this fence was heavily concentrated in lower Michigan, New Jersey, northern Pennsylvania, New York, and northern New England, although it existed throughout most of the eastern United States; Quebec.
Variant. Two posts are used instead of one at each end of each set of rails, which are nailed or wired in place.

BOARD FENCE
This fence became popular in the later nineteenth century after the advent of dimension lumber and low-cost nails. Posts were sometimes square lumber lengths, although timber continued to be used. Usually four horizontal boards were nailed to the posts. Used to restrain animals but expensive to construct, this fence is typically found only in horse-raising areas.

Fig. 13.9 The post-and-rail fence.

Range: Extensive in Bluegrass Basin, Nashville Basin, northern Virginia; scattered elsewhere.

JACAL FENCE
In western areas of little timber, any long, thin local material, such as the cactus called ocotillo, is often used. The light material is laid horizontally and held between split upright posts fastened together with wire to form a palisade.
Range: Desert and near desert Southwest of the United States.

BARBED-WIRE FENCE
Twisted strands of galvanized iron wire with attached barb form this common fence. Douglas Leechman reports, in "Good Fences Make Good Neighbors," that "Probably no other single invention of the nineteenth century had such a profound effect on the lives of people on the land as did barbed wire" (p. 233). An early center of barbed-wire modification and invention was in north-central Illinois. Great variety of styles.
Range: Ubiquitous in North America.
Variant. In the Smokey Hills region of Kansas, the barbed-wire strands (and some woven wire too) are carried on squared-off limestone fence posts. Such posts were a durable substitute for wood, which was more expensive and generally not plentiful on the mostly treeless plains.

Fig. 13.10 A board fence, because of its maintainence expense, is normally used only to restrict animals and not to enclose fields. This fence is in northern Summit County, Ohio.

JACK FENCE

Like the Russell fence, the jack fence is suitable for those areas where postholes cannot be dug easily. The fence consists of a triangular jack in which two poles are joined at the top and butt against the ground, sometimes with a low cross-member for support. The jacks support rails that are spiked or wired to the jacks.

Fig. 13.11 The jacal fence.

Fig. 13.12 A barbed-wire fence is the most common North American rural fence.

Range: Similar to that of the Russell fence, but wider to include Canadian Rockies and the Great Basin.

Variant. The western snow fence uses jacks of dimension lumber and wide, sawn boards instead of rails, which may be horizontal or vertical.

MORMON FENCE

Like the Mormon hay derrick, this landscape feature is so strongly associated with Mormon settlement that it is diagnostic. Perhaps a result of the isolation and poverty of the early Mormon settlements in the West or perhaps due to frugality, the Mormon fence is a hodgepodge of whatever is available. The fence consists of pickets attached to upper and lower horizontal rails, in turn nailed to widely spaced posts. The pickets are often scrap wood of various sizes, including worn-out tools and old wagon wheels. These fences are never painted, whitewashed, or stained.

Range: Utah, and Mormon settlements in surrounding states.

WOVEN-WIRE FENCE

This fence was introduced in the early 1880s and adopted especially by sheep raisers. Unlike barbed wire, woven-wire fences were coyote-, wolf-, and even rabbit-proof, if the mesh was small enough. Six-inch mesh, 42–52 inches high, was strung between cedar posts. Often a strand of barbed wire was run along the ground or three strands were placed throughout the woven wire. Frequently in the West, the top strand of barbed wire served as the ranch telephone line.

Other term: Net-wire.

Range: Widespread, especially throughout the Midwest and West, often along railroad and highway rights-of-way.

ELECTRIC FENCE

This fence usually consists of a single strand of wire, but sometimes is used with other wire fences. One six-volt battery can charge a fence five or six miles long. Highly visible white porcelain insulators mark older fences. This fence may have been introduced on the Texas prairies in the closing years of the nineteenth century, but it was not widely adopted until the early 1930s.

Range: Widespread, to restrain cattle.

Fig. 13.13 A woven-wire fence located in Summit County, Ohio.

Fig. 13.14 The electric fence is easily identified by the white insulators.

Fig. 13.15 This stile is located on an Amish farm in Holmes County, Ohio. Several other kinds of stiles exist.

RUBBER-STRIP FENCE

A new alternative to expensive wire fences is made from rubber tires, cut and vulcanized into long thin strips and stretched across fence posts.

Range: Eastern Midwest, especially in Ottawa and Allegan counties, Michigan.

STILES

Stiles usually consist of a wooden framework that offers a means to climb up and over a fence. The stile straddles the fence, so that one side is a duplicate of the other. Because stiles are small and inconspicuous, they are often hard to find.

Range: Eastern United States and Canada, especially prevalent in Amish settlement areas.

P A R T   T H R E E

Where and What

Part 3 is designed to help readers discover the structural treasures of the rural countryside, which unfortunately often so resemble one another as to be confusing. "Where to Look, and for What" lists the structures in alphabetical order and some of the most likely areas where they can be found. Persons looking for particular barns or other structures will find this section helpful.

"What to Look For, and Where" lists geographical areas in alphabetical order, with an indication of the structures that frequently occur in these areas. The reader should remember that not every feature is listed under the various geographical headings. Those listed are often only the most commonly occurring, or the unique ones.

Where to Look, and for What

This list does not include all places where the various features can be found, only some of the most likely. Remember that certain features are difficult to find even in the most likely place.

Acadian barn	Madawaska and St. John valleys of New Brunswick/Maine
Amish barn	Madison County, Ohio
Appalachian barn	throughout central and southern Appalachia; Missouri and Arkansas
Appalachian Meadow barn	southeastern Ohio; West Virginia
Arched roof	see "Gothic roof"
Bakeoven, Belgian	Door and Kewaunee counties, Wisconsin
Bakeoven, French Canadian	Quebec; New Brunswick; Nova Scotia
Bakeoven, German	Pennsylvania, especially Bucks County
Bakeoven, Spanish-Mexican	southwestern United States
Basement Drive-Through barn	south-central Pennsylvania; central Maryland; Shenandoah valley, Virginia
Beaverslide hay stacker	see "Hay stackers"
Bluegrass barn	see Transverse Frame barn, Bluegrass variant
Brick barns	south-central Pennsylvania; north-central Maryland; elsewhere in areas of shale rock
Cajun barn	French areas of Louisiana
Cantilevered Double-Crib barn	eastern Tennessee; western North Carolina; eastern Kentucky

Cellars	Great Plains and Midwest; southern Ontario
Cisterns, Cajun	Louisiana
Closed-Forebay Standard barn	Pennsylvania; central Maryland; Wisconsin
Cobblestone walls and foundation	western and central New York; around Paris, Ontario
Connecticut barn	see "English barn"
Cordwood walls	see "Stovewood walls"
Corncrib, Circular	Midwest; eastern United States; Ontario
Corncrib, Drive-in	Midwest; eastern United States; Ontario
Corncrib, Gable-Roofed	eastern United States and Canada
Corncrib, Midwest Masonry	Iowa; Illinois, and less frequently, elsewhere in the Corn Belt
Corncrib, Shed-Roofed	Midwest; eastern United States; Ontario
Corncrib, Slant-Sided	eastern Midwest; New England; Ontario
Cowsheds	northern New Mexico
Crib barns	Appalachia, especially southern part; Canadian Shield; Missouri and Arkansas; mountainous western United States and Canada
Czech barn	southeastern South Dakota
Decorated Brick-End barns	south-central Pennsylvania; north-central Maryland
Decorated doors	southern Michigan; northeastern Indiana; northwestern Ohio; lower Mohawk and Schoharie valleys of New York
Decorated roofs	eastern Midwest, especially around Rochester, Indiana
Diamond-shaped gable window	Vermont; upstate New York; southern Michigan; Wisconsin; southern Ontario
Domestic tankhouses	California; Oregon and Washington; central Great Plains
Double-Crib barn (Type 1)	Appalachia and Deep South

Double-Crib barn (Type 2)	Middle Atlantic states
Double-Crib barn (Type 3)	hilly areas of eastern half of United States
Double-Crib barn (Type 4)	Calhoun County, Illinois; Little Dixie area of Missouri
Double-Decker barn	York County, Pennsylvania; Midwest
Drive-in Crib barn	wide distribution from Virginia and North Carolina to Kansas and Nebraska
Dryhouses	Pennsylvania
Dutch barn	eastern New York; northern New Jersey; southwestern Michigan
English barn	southern New England; New York; New Jersey; eastern Midwest; Utah
English Bank barn	from New York to eastern Midwest, also Wisconsin
Entry porches	east-central New York; from southwestern New York across Pennsylvania panhandle to northeastern Ohio
Erie Shore barn	lower Great Lakes; southern Ontario; Puget Sound lowland of Washington
Extended Pennsylvania barn	southeastern Pennsylvania
Fan windows	see "Window, half-round"
Feeder barn	throughout the Midwest; West Coast from northern California to British Columbia
Fence, barbed-wire	ubiquitous in North America
Fence, board	Bluegrass Basin; Nashville Basin; northern Virginia
Fence, brush	northern Michigan
Fence, electric	widespread, wherever cattle are grazed.
Fence, Irish	northern Arizona; southern Appalachia
Fence, jacal	desert and near-desert Southwest
Fence, jack	similar to the Russell fence, but wider distribution to include Canadian Rockies and the Great Basin
Fence, log-and-chock	now mostly found in mountain basins of western Canada and United States
Fence, Mormon	Utah, and Mormon settlements in surrounding states

Fence, post-and-rail	lower Michigan; New Jersey; northern Pennsylvania; New York and northern New England; less common in most of the rest of eastern United States
Fence, rail	No other fence was so widely distributed throughout the entire United States and Canada in the nineteenth century before the introduction of wire fencing. Still common in central Ontario, central Pennsylvania, western Virginia, and generally throughout Appalachia
Fence, rubber-strip	eastern Midwest
Fence, Russell	interior areas of Pacific Northwest and British Columbia
Fence, sawyer	very scattered locations
Fence, Shanghai	see "Fence, Irish"
Fence, stake-and-rider	southern part of Canadian Shield; north-central Pennsylvania; central Appalachia
Fence, stone	northern Great Lakes and areas of glacial deposition in northeastern North America
Fence, stump	very scattered, but usually where forest cover is prevalent
Fence, woven-Wire	widespread, especially throughout the Midwest and West, often along railroad and highway rights-of-way
Fieldstone barns	glaciated areas of United States and Canada, southeastern Wisconsin; Hamilton County, Ohio.
Finnish barn	upper Great Lakes; Canadian Shield
Foundation, brick	sparse distribution, but usually in areas of shale rock
Foundation, cutstone	areas of limestone or sandstone rock outcrops
Foundation, fieldstone	glaciated areas of United States and Canada
Four-Crib barn	Cumberland basin of north-central Tennessee and south-central Kentucky
Front-Drive Crib barn	south and central Appalachia
Gable-Entry Bank barn	western New England; central New York; north-central Pennsylvania
Gable roof	throughout United States and Canada

Gambrel roof	throughout United States and Canada; especially common in lower Great Lakes and eastern Midwest
German barn	see "Standard Pennsylvania barn"
Gothic roof	most common in Midwest
Granaries	eastern and central North America
Grundscheier	southeastern Pennsylvania
Grundscheier Type D	southeastern Pennsylvania; southern Wisconsin
Grundscheier with Forebay	Bedford, York, and Adams counties, Pennsylvania; from Virginia/West Virginia border to eastern Kentucky; eastern Tennessee; western North Carolina
Hanging gable	southern Midwest; Great Plains; southeastern United States; sparsely scattered over entire East Coast
Hay barracks	very rare, scattered in eastern Massachusetts, Virginia, Maryland, Ohio, eastern Iowa, northern Illinois, western New York, Rhode Island, Prince Edward Island, southeastern Manitoba, Irish farms in eastern Newfoundland; especially Hudson valley and central New Jersey; western Wisconsin
Hay derricks	Utah, and Mormon settlement areas in surrounding states
Hay hoods	Appalachia; southern Midwest; interior valleys of Pacific Coast (Willamette valley of Oregon, for example)
Hay stackers	western North America, especially Montana, Wyoming, and Alberta
Hay stacker, Beaverslide	especially Montana, Wyoming, Colorado, and Utah
Hay stacker, combination	Great Plains and Prairie provinces
Hay stacker, overshot	Great Plains and Prairie provinces
Hay stacker, ram	throughout the northern Great Plains and southern Prairie provinces
Hay Stacker, swinging	Great Plains and Prairie provinces

Hispanic Twin-Crib barn	northern New Mexico
Hop barn	central New York; southern Wisconsin; northern California; Willamette valley, Oregon
Limestone barns	Shenandoah valley of Virginia; Hagerstown valley of Maryland; Cumberland, Lebanon, and Lehigh valleys of Pennsylvania; Flint Hills of Kansas; the Edwards Plateau of Texas; elsewhere in local areas of limestone bedrock
Log barns	Appalachia; Canadian Shield; Missouri and Arkansas; eastern Texas; northern or upper Great Lakes; Rocky Mountains; Utah and Nevada
Madawaska Twin barn	St. John valley, Maine; southern Quebec; New York
Manitoba Mennonite barn	southeastern Manitoba
Martin holes	see "Owl holes"
Midwest Three-Portal barn	western Kentucky; southern Indiana; Illinois; central Midwest
Milkhouses	northeastern United States and adjacent Canada
Monitor roof	interior valleys of California; Winooski valley of Vermont
Mormon barn	see "English barn"
Mormon Thatched-Roof cowshed	see "Cowsheds"
Mountain Horse barn	northern and central Rocky Mountains
Multiple-Overhang Standard barn	Shenandoah, Rockingham, and Augusta counties, Virginia; Perry, Fairfield, Allen, and Putnam counties, Ohio; Onondaga County, New York; Green County, Indiana
Murals, barn wall	southern Wisconsin; southeastern Michigan
New England barn	another term for English barn, not to be confused with New England Connected barn
New England Connected barn	southern Maine; Vermont; New Hampshire; Massachusetts; extreme eastern New York

Open-Forebay standard barn	central Pennsylvania to Wisconsin
Owl holes	southeastern Ohio; Pennsylvania; New Jersey; southern Ontario
Outshed Barns	southeastern and central Pennsylvania; northern Maryland
Pentice (roof)	Middle Atlantic states
Pent roof (on barn gable-end)	northwestern Ohio, northern Indiana and southern Michigan, in areas of original English settlement; east-central New York
Pent roof (on barn side)	middle Atlantic states and eastern Midwest in areas of original German settlement; southeastern Wisconsin
Pole barns	widespread, but especially common in the Midwest
Polygonal and Round barns	Ozaukee County, Wisconsin; Stephenson County, Illinois; Fulton County, Indiana; widely scattered elsewhere
Posted-Forebay Standard barn	Pennsylvania to Wisconsin; Shenandoah valley of Virginia; southern Ontario
Potato barn	Aroostook County, Maine; central Wisconsin; Willamette valley, Oregon; Snake River plain, Idaho
Privies	everywhere, but becoming harder to find
Quebec Long barn	French Canada
Rack-sides	Pennyroyal section of Kentucky
Rainbow roof	see "Gothic roof"
Raised barn	New York; southern Ontario; upper Midwest
Ramp-Shed barn	Pennsylvania; Ohio (especially Fulton County)
Round roof	see "Gothic roof"
Saltbox roof	widespread, especially found in Dutch areas of southwestern Michigan (Allegan and Ottawa counties)
Scale houses	Pike County, Ohio; scattered widely in the Corn Belt
Scioto Valley Elevated corncrib	Pike County, Ohio
Side-Drive crib	Appalachia, especially in Kentucky and Tennessee

Siding, horizontal	New England; eastern New York; Shenandoah valley of Virginia; northern Wisconsin
Silo, Cement-Stave	widely diffused through the United States
Silo, Fieldstone	mostly in glaciated areas of northeastern United States
Silo, Harvestore	nearly ubiquitous in eastern North America
Silo, Horizontal	common only on the Great Plains, but also scattered throughout the Dairy Belt
Silo, Masonry	throughout the Dairy Belt, but especially in Wisconsin
Silo, Octagonal	rare but widespread; small concentration in Vermont
Silo, Poured-Concrete	entire Dairy Belt
Silo, Rectangular Wooden	northeast Ohio; southwestern New York; Puget Sound; Willamette valley of Oregon; scattered elsewhere
Silo, Wooden-Hoop	rare but widespread; small concentration in eastern Wisconsin
Silo, Wooden-Stave	common throughout the eastern Dairy Belt
Slab wall	northeastern Pennsylvania
Smokehouse, Cantilevered-Roof	scattered through Appalachia from Pennsylvania to east Texas
Smokehouse, Dutch	New Jersey; eastern New York
Smokehouse, English	tidewater Virginia, Maryland, New Jersey
Smokehouse, Pennsylvania German	eastern Pennsylvania; scattered elsewhere in places of German settlement
Smokehouse, Two-Story	scattered through Appalachia from Virginia to east Texas
Smokehouse, Upland South	central and southern Appalachia
Sorghum mills	Appalachia; Ozarks; southern Wisconsin
Southern Ontario barn	see "Raised barn"
Springhouses	entire continent
Stackwood walls	see "Stovewood walls"
Standard Pennsylvania barn	from eastern Pennsylvania to the central Midwest; Virginia; southern Ontario

Stiles	hard to find; look in Amish areas
Stone fence posts	Smokey Hills of central Kansas
Stone wall, Mortared	occasional in areas of glacial deposition
Stone wall, stacked	limestone areas, including Bluegrass Basin of Kentucky; eastern Wisconsin; Texas Hill Country
Storm cellar	Midwest
Stovewood walls	upper Great Lakes, especially Delta County, Michigan
Sugarhouses	from northeastern North America westward into the upper Great Lakes; especially common in New England; eastern Ontario, Quebec; New Brunswick; northern New York; northeastern Ohio
Summer kitchens	Midwest; Pennsylvania; Door County, Wisconsin
Swallow holes	see "Owl holes"
Swedish barn	Chisago County, Minnesota; central Wisconsin
Sweitzer barn	Pennsylvania to Missouri; southern Ontario in German settled areas; Shenandoah valley of Virginia
Sweitzer Half barn	central Maryland; southeastern Wisconsin
Tapeista	northern New Mexico
Tasolera	northern New Mexico
Three-Bay Threshing barn	see "English barn"
Three-End barn	eastern Wisconsin; Columbiana County, Ohio
Timber frame barns	throughout United States and Canada, but rare on Great Plains
Tobacco barns	Kentucky; Tennessee; central and southern Virginia, North Carolina; southern Ontario; south-central Wisconsin, southwestern Ohio, Vernon County, Wisconsin, Sprague County, Minnesota
Transverse Frame barn	central and southern Appalachia; southern and central Midwest; southwestern Wisconsin
Transverse Frame barn, Bluegrass variant	Bluegrass Basin of Kentucky

Up-Country Posted-Forebay barn	Pennsylvania to Virginia, westward to Wisconsin
Ventilators, covered gable	Lancaster County, Pennsylvania; west-central Ohio
Ventilators, Cupola ridge	dairy belt of northern Midwest; northeastern United States; eastern Canada
Ventilators, open gable	central and southern Appalachia
Ventilators, slit	eastern Pennsylvania
Walls, cedar or pine shake	coastal New England; New Jersey
Welsh barn	see "Gable-Entry Banked barn"
Windmills	from New England to California; especially found on Great Plains and Prairie provinces; often indicates Amish settlements
Window, half-round	eastern Pennsylvania; Lake Michigan shoreline of Wisconsin
Window, porthole	Door Peninsula of Wisconsin; south-central and eastern Wisconsin
Wisconsin Dairy barn	Minnesota to New England in dairy-farming areas; east of Buhl, Idaho
Yankee barn	see "English barn"

What to Look For, and Where

This list indicates structures that are especially important or common in the places mentioned. In some instances, identical or similar structures can also be found in locations not listed. Often a location is listed because studies have been published for that area or the authors are familiar with it. Consult the previous section for additional possibilities not given here.

Alberta	hay stackers
Arizona	bakeovens, jacal fences
Arizona (northern)	Irish fences
Appalachia	log barns, hay hoods, crib barns, open-gable ventilators, Front-Drive cribs, Side-Drive cribs, Appalachian barns, Transverse Frame barns, hanging gables, smokehouses, sorghum mills, rail fences, stake-and-rider fences, Irish fences
British Columbia	Russell fences
California (interior)	monitor roofs, Feeder barns, domestic tankhouses
California (northern)	hop barns, Transverse Frame barns, domestic tankhouses
Canada (French)	Quebec Long barn, bakeovens
Canadian Shield	log barns, crib barns, Finnish barns, rail fences, stake-and-rider fences
Great Plains	hanging gables, Feeder barns, windmills, domestic tankhouses, storm cellars, horizontal silos, hay stackers
Idaho	hay stackers, Mormon hay derricks, Mountain Horse barns

Idaho (Snake River Plain)	Potato barns, hay derricks, hay stackers, Wisconsin Dairy barns (east of Buhl)
Illinois (Calhoun County)	Double-Crib barn (Type 4)
Illinois (Stephenson County)	Polygonal and Round barns
Illinois (western)	Midwest Masonry corncribs, Feeder barns
Indiana (Green County)	Multiple-Overhang Standard barns
Indiana (northeast)	gable-end pent roofs, decorated doors, milk houses, English Bank barns
Indiana (southern)	Midwest Three-Portal barns
Iowa	Midwest Masonry corncribs
Kansas (Flint Hills)	limestone barns, windmills
Kansas (Smokey Hills)	limestone fence posts
Kentucky	Transverse Frame barns, crib barns, log barns, Midwest Three-Portal barn, Tobacco barns, Side-Drive cribs
Kentucky (Bluegrass Basin)	stacked stone walls, board fences
Kentucky (Pennyroyal section)	rack-sides
Kentucky (south central)	Four-Crib barns, Tobacco barns
Louisiana	Cajun barns, Cajun cisterns, bakeovens
Lower Great Lakes	gambrel roofs, Erie Shore barns, milk houses
Maine (Aroostook County)	Potato barns
Maine (southern)	New England Connected barn, English barns, milk houses, sugarhouses, post-and-rail fences
Maine (St. John valley)	Madawaska Twin barns, milk houses, Acadian barns
Manitoba (southeast)	Mennonite housebarns, hay stackers
Maryland	smokehouses

Maryland (central)	Closed-Forebay Standard barn, Basement Drive-Through barn, Up-Country Posted-Forebay barns, milk houses
Maryland (Hagerstown valley)	limestone barns
Maryland (north central)	brick barns, Decorated Brick-End barns, outshed barns, Closed-Forebay Standard barns, Sweitzer Half barns, Basement Drive-Through barns, milk houses
Massachusetts	English barn, horizontal siding, Gable-Entry Bank barn, New England Connected barn, milk houses, Slant-Sided corncribs, post-and-rail fences
Michigan (Allegan and Ottawa counties)	saltbox roofs, Dutch barns, diamond-shaped gable window
Michigan (northern)	brush fences
Michigan (southeast)	barn wall murals, Raised barns, post-and-rail fences, milk houses
Michigan (southern)	gable-end pent roofs, diamond-shaped gable windows, decorated doors, milk houses, post-and-rail fences
Midwest	Midwest Three-Portal barns, English barns, English Bank barns, Raised barns, gambrel roofs, Gothic roofs, sidewall pent roof, hay hoods, hanging gables, decorated roofs, Drive-in cribs, Transverse Frame barns, Sweitzer barns, Up-Country Posted-Forebay barns, Double-Decker barns, cupola ridge ventilators, Open-Forebay Standard barns, Pole barns, storm cellars, Slant-Sided corncribs, Shed-Roofed corncribs, Drive-in corncribs, Circular corncribs, rubber-strip fences
Minnesota	Wisconsin Dairy barns, Gothic roofs, Raised barns, Multiple-Overhang Standard barns, Basement Drive-Through barns, milk houses

Minnesota (Chisago County)	Swedish barns
Missouri (central)	Stone-Arch Forebay barns, sorghum mills
Montana	hay stackers
Nebraska (south-eastern)	Stone-Arch Forebay barns, windmills
Nevada	log barns, Mormon Thatched-Roof cow-sheds
New Brunswick	smokehouses, sugarhouses, Acadian barns
New England	horizontal siding, English barns, sugar-houses, Slant-Sided corncribs, hay bar-racks
New England (coastal)	walls of cedar or pine shakes
New Hampshire	New England Connected barn, English barns, milk houses, sugarhouses, post-and-rail fences
New Jersey	owl holes, English barns, Dutch barns, smokehouses, hay barracks, post-and-rail fences
New Jersey (Monmouth and Somerset counties)	walls of cedar or pine shakes
New Mexico	bakeovens, jacal fences
New Mexico, (northern)	Tasolera, Tapeista, Hispanic Twin-Crib barns
New York (central)	Raised barns, Gable-Entry Bank barns, entry porches, diamond-shaped gable windows, Hop barns, milk houses, sug-arhouses, Slant-Sided corncribs, cob-blestone walls and foundations
New York (eastern)	horizontal siding, Dutch barns, gable-end pent roofs, entry porches, diamond-shaped gable windows, decorated doors, English barns, Raised barns, New Eng-land Connected barns, milk houses, smokehouses, sugarhouses, Slant-Sided corncribs
New York (Hudson valley)	Dutch barns, hay barracks

New York (Mohawk valley)	decorated doors, Raised barns, Dutch barns, milk houses
New York (Onondaga County)	Multiple-Overhang Standard barns, milk houses
New York (Schoharie valley)	decorated doors, Raised barns, Dutch barns, hay barracks
New York (southwestern)	entry porches, rectangular silos
North Carolina (western)	log barns, crib barns, Grundscheier with Forebay, Tobacco barns, Cantilevered Double-Crib barns
Nova Scotia	smokehouses
Ohio (eastern)	Up-Country Posted-Forebay barns, Standard Pennsylvania barn, entry porches
Ohio (Fulton County)	Ramp-Shed barns, twin barns
Ohio (Hamilton County)	fieldstone barns
Ohio (Madison County)	Amish barns, windmills
Ohio (northeast)	entry porches, English barns, milk houses, rectangular silos, sugarhouses
Ohio (northwest)	gable-end pent roofs, decorated doors, English barns
Ohio (Scioto valley)	elevated corncribs, scale houses
Ohio (southeast)	owl holes, Sweitzer barns, Appalachian Meadow barns
Ohio (southwest)	Tobacco barns
Ohio (west central)	covered-gable ventilators, Multiple-Overhang Standard barns
Ontario (eastern)	sugarhouses
Ontario (southern)	Raised barns, Erie Shore barns, Sweitzer barns, diamond-shaped gable windows, metal cupola ridge ventilators, Posted-Forebay Standard barn, Tobacco barns, Standard Pennsylvania barn, Wisconsin Dairy barns, milk houses, sugarhouses, Slant-Sided corncribs, Shed-Roofed corncribs, Drive-in corncribs, Circular corncribs, post-and-rail fences, owl holes

Oregon (Willamette valley)	hay hoods, Feeder barns, Potato barns, Hop barns, milk houses, domestic tankhouses, rectangular silos
Ozarks	log barns, crib barns, Appalachian barns, sorghum mills
Pennsylvania	owl holes, Standard Pennsylvania barns, Sweitzer barns, Up-Country Posted-Forebay barns, Closed-Forebay Standard barn, Open-Forebay Standard barn, Ramp-Shed barns, milk houses, dry houses
Pennsylvania (Cumberland valley)	limestone barns
Pennsylvania (eastern)	slit ventilators, Standard Pennsylvania barns, Grundscheier, Extended Pennsylvania barns, Grundscheier Type D, outshed barns, half-round windows, milk houses, smokehouses, bakeovens
Pennsylvania (Lancaster County)	covered-gable ventilators, Standard Pennsylvania barn, Transitional Sweitzer, windmills, bakeovens, stiles
Pennsylvania (Lebanon valley)	limestone barns, smokehouses
Pennsylvania (Lehigh valley)	limestone barns
Pennsylvania (Montgomery and Lehigh counties	Stone-Arch Forebay barns
Pennsylvania (north central)	gable-Entry Banked barn, milk houses, rail fences, stake-and-rider fences, post-and-rail fences
Pennsylvania (northeastern)	slab stone walls
Pennsylvania (northwest)	entry porches, milk houses, sugarhouses, Raised barns
Pennsylvania (south central)	Brick barns, Decorated Brick-End barns, Standard Pennsylvania barns, Sweitzer barns, Grundscheier with Forebay barns, Up-Country Posted-Forebay barns, outshed barns, Closed-Forebay Standard barns, Open-Forebay Standard barns, Basement Drive-Through barns, milk houses

Pennsylvania (southeast)	Grundscheier, Grundscheier Type D, Sweitzer, log Sweitzer, Transitional Sweitzer Extended Pennsylvania barns, Up-Country Posted-Forebay barns, summer kitchens
Pennsylvania (York County)	Double-Decker barns, Grundscheier
Prince Edward Island	hay barracks
Quebec	smokehouses, sugarhouses
Rocky Mountains	log barns, crib barns, log-and-chock fences, jack fences, Mountain Horse barns, hay stackers
South Dakota (southeast)	Czech barns, windmills
Tennessee (eastern)	crib barns, Grundscheier with Forebay barns, log barns, Cantilevered Double-Crib barns
Tennessee (north central)	four-Crib barns, Tobacco barns
Texas (eastern)	smokehouses, domestic tankhouses
Texas (Edwards Plateau)	limestone barns, stacked stone walls
Upper Great Lakes	log barns, Finnish barns, stovewood walls, sugarhouses, rail fences
Utah	log barns, English barns (Mormon barns), hay derricks, Mormon fences, Mormon Thatched-roof cowsheds
Vermont	diamond-shaped gable windows, New England Connected barns, Wisconsin Dairy barns, Raised barns, Gable-Entry Bank barns, milk houses, sugarhouses, Octagonal silos, post-and-rail fences
Vermont (northeastern)	entry porches
Vermont (northern)	barn murals
Vermont (southeast)	Dutch barns
Vermont (Winooski valley)	monitor roofs
Virginia	Up-Country Posted-Forebay barns, Standard Pennsylvania barns, Grundscheier with Forebay barns

Virginia (northern)	board fences
Virginia (Shenandoah valley)	limestone barns, horizontal siding, Up-Country Posted-Forebay barns, Sweitzer barns, Posted-Forebay Standard barns, Multiple-Overhang Standard barns, Basement Drive-Through barns
Virginia (Tidewater)	smokehouses
Washington	domestic tankhouses, Transverse Frame barns, Russell fences, rectangular silos
West Virginia	Appalachian Meadow barns, crib barns, Grundscheier with Forebay barns, rail fences
Wisconsin	Closed-Forebay Standard barn, diamond-shaped gable window, Open-Forebay Standard barns, Posted-Forebay Standard barns, Up-Country Posted-Forebay barns, Wisconsin Dairy barns, milk houses, Masonry silos
Wisconsin (central)	Up-Country Posted-Forebay barns (Wisconsin Porch barn), Tobacco barns, Potato barns, milk houses, Masonry silos, Swedish barns
Wisconsin (Door Peninsula)	log barns, porthole gable windows, stovewood walls, bakeovens, summer kitchens, privies next to bakeovens and summer kitchens
Wisconsin (eastern)	Three-End barns, English barns, milk houses, Wooden-Hoop silos, Masonry silos
Wisconsin (Ozaukee County)	Polygonal and Round barns, milk houses, Up-Country Posted-Forebay barns
Wisconsin (southern)	barn wall murals, Raised barns, Grundscheier Type D, Hop barns, milk houses, sorghum mills
Wisconsin (southeastern)	Sweitzer Half barn, sidewall pent roof, milk houses
Wisconsin (southwest)	Transverse Frame barns, Midwest Three-Portal barns
Wyoming	hay stackers

Sources for More Information

For readers who wish more detailed information on certain barn types, or on the various barn and farmyard features, the most comprehensive volume currently available is Allen G. Noble, *Barns and Other Farm Structures*, volume 2 of *Wood, Brick and Stone: The North American Settlement Landscape* (Amherst: University of Massachusetts Press, 1984).

The following sources are listed by barn type, particular feature, or specific farm structure.

Barn Bibliographies

CALKINS, CHARLES F. *The Barn as an Element in the Cultural Landscape of North America: A Bibliography*. Monticello, Ill.: Vance Bibliographies, 1979.

CARLSON, ALVAR W. "Bibliography on Barns in the United States and Canada." *Pioneer America*, vol. 10, no. 1 (1978).

SHULTZ, LeROY G. *Barns, Stables and Outbuildings*. Jefferson, N.C.: McFarland, 1986. Must be used with caution since many enteries are incorrect or incomplete.

Acadian Barns

KONRAD, VICTOR A. "Against the Tide: French Canadian Barn Building in the St. John Valley of Maine." *American Review of Canadian Studies*, vol. 12, no. 2 (Summer 1982).

Amish Barns

WILHELM, HUBERT G. H. "Amish-Mennonite Barns in Madison County, Ohio: The Persistence of Traditional Form Elements." *Ohio Geographers: Recent Research Themes*, vol. 4 (1976).

Bake Ovens

BOILY, LISE, and JEAN-FRANÇOIS BLANCHETE. *The Breadovens of Quebec*. Ottawa: National Museums of Canada, 1979.

CALKINS, CHARLES, and WILLIAM LAATSH. "The Belgian Outdoor Ovens of Northeastern Wisconsin." *P.A.S.T.: Pioneer America Society Transactions,* vol. 2 (1979).

KNIFFEN, FRED B. "The Outdoor Oven in Louisiana." *Louisiana History,* vol. 1 (1960).

LESSARD, MICHEL, and GILLES VILANDRE. In *La maison quebecoise,* 255–64. Montreal: Les Editions de l'Homme, 1974.

LONG, AMOS. "Bakeovens in the Pennsylvania Folk-Culture." *Pennsylvania Folklife,* vol. 14, no. 2 (December 1964).

MINDELEFF, VICTOR *A Study of Pueblo Architecture: Tusayan and Cibola.* Eighth Annual Report of the Bureau of American Ethnology. Washington, D.C.: Smithsonian Institution, 1891.

Barn Decoration

BECK, ROBERT L., and GEORGE W. WEBB. "Decoration in the Boundary Junction Area of Indiana, Michigan and Ohio: The Case of the Painted Arch." Professional Paper No. 9. Department of Geography and Geology, Indiana State University, Terre Haute, 1977.

MAHR, AUGUST C. "Origin and Significance of Pennsylvania Dutch Barn Symbols." *Ohio State Archaeological and Historical Quarterly,* vol. 54 (1945).

STOUDT, J. *Decorated Barns of East Pennsylvania.* Plymouth, Pa.: Plymouth Meeting, 1945.

Barns, Regional or Local

APPS, JERRY and ALLEN STRANG. *Barns of Wisconsin.* Madison, Wisc.: Tamarack Press, 1977.

DANDEKAR, HEMALATA, and DANIEL F. SCHOOF. "Michigan Farms and Farm Buildings: 150 Years of Transformation." *Inland Architect,* vol. 32, no. 1 (January/February 1988).

ENNALS, PETER M. "Nineteenth-Century Barns in Southern Ontario." *Canadian Geographer,* vol. 16, no. 3 (Fall 1972).

GRITZNER, CHARLES F. "Log Barns of Hispanic New Mexico." *Journal of Cultural Geography,* vol. 10, no. 2 (Spring/Summer 1990).

HARTMAN, LEE. "Michigan Barns: Our Vanishing Landmarks." *Michigan Natural Resources,* vol. 45 (1976).

JACKSON, J. B. "A Catalog of New Mexico Farm Building Terms." *Landscape,* vol. 1, no. 3 (Winter 1952).

KILPINEN, JON T. "The Mountain Horse Barn: A Case of Western Innovation." *P.A.S.T: Pioneer America Society Transactions,* vol. 17 (1994).

KRUCKMAN, LAURENCE, and DARRELL L. WHITEMAN. "Barns, Buildings and Windmills: A Key to Change on the Illinois Prairie." *Journal of the Illinois State Historical Society,* vol. 68 (1975).

McHENRY, STEWART. "Vermont Barns: A Cultural Landscape Analysis." *Vermont History,* vol. 46 (Summer 1978).

NOBLE, ALLEN G. "Barns and Square Silos in Northeastern Ohio." *Pioneer America,* vol. 6, no. 2 (July 1974).

SCULLE, KEITH A., and H. WAYNE PRICE. "The Traditional Barns of Hardin County, Illinois: A Survey and Interpretation." *Material Culture,* vol. 25, no. 1 (Spring 1993).

Brick-End Barns
STAIR, J. WILLIAM. "Brick-End Decorations." In Alfred L. Shoemaker, ed., *The Pennsylvania Barn.* Lancaster, Pa.: Pennsylvania Dutch Folklore Center, 1955.

Cajun Barns
COMEAUX, MALCOLM L. "The Cajun Barn." *Geographical Review,* vol. 79, no. 1 (January 1989).

Cantilevered Double-Crib Barns
MOFFETT, MARIAN, and LAWRENCE WODEHOUSE. *East Tennessee Cantilever Barns.* Knoxville: University of Tennessee Press, 1993.

Cellars
ROARK, MICHAEL. "Storm Cellars: Imprint of Fear on the Landscape." *Material Culture,* vol. 24, no. 2 (Summer 1992).

Chicken Houses
LONG, AMOS, JR. "The Pennsylvania German-Family Farm." *Pennsylvania German Society,* vol. 6 (1972).
PASSARELLO, JOHN. "Adaptation of House Type to Changing Functions: A Sequence of Chicken House Styles in Petaluma." *California Geographer,* vol. 5 (1964).

Corncribs
ROE, KEITH E. *Corncribs in History, Folklife and Architecture.* Ames: Iowa State University Press, 1988.
SCHIMMER, JAMES R., and ALLEN G. NOBLE. "The Evolution of the Corn Crib with Special Reference to Putnam County, Illinois." *P.A.S.T: Pioneer America Society Transactions,* vol. 7 (1984).
SCHULTZ, LEROY. "West Virginia Cribs and Granaries." *Goldenseal,* vol. 9, no. 4 (Winter 1983).

Cowsheds (see Mormon Cowsheds)

Crib Barns
GLASSIE, HENRY. "The Double Crib Barn in South Central Pennsylvania, Part 1." *Pioneer America,* vol. 1, no. 1 (January 1969).
———. "The Double Crib Barn in South Central Pennsylvania, Part 2." *Pioneer America,* vol. 1, no. 2 (July 1969).

———. "The Double Crib Barn in South Central Pennsylvania, Part 3." *Pioneer America*, vol. 2, no. 1 (January 1970).

———. "The Double Crib Barn in South Central Pennsylvania, Part 4." *Pioneer America*, vol. 2, no. 2 (July 1970).

———. "The Old Barns of Appalachia." *Mountain Life and Work*, vol. 45 (Summer 1965).

JORDAN, TERRY G., and MATTI KAUPS. *The American Backwoods Frontier: An Ethnic and Ecological Interpretation.* Baltimore: Johns Hopkins University Press, 1989.

MONTELL, WILLIAM LYNWOOD, and MICHAEL LYNN MORSE, *Kentucky Folk Architecture.* Lexington: University Press of Kentucky, 1976.

MORGAN, JOHN, and ASHBY LYNCH, JR. "The Log Barns of Blount County, Tennessee." *Tennessee Anthropologist*, vol. 9, no. 2 (Fall 1984).

PRICE, H. WAYNE. "The Double-Crib Log Barns of Calhoun County." *Journal of the Illinois State Historical Society*, vol. 73, no. 2 (Summer 1980).

Czech Barns

RAU, JOHN E. "Czechs in South Dakota." In Allen G. Noble, ed., *To Build in a New Land*. Baltimore: Johns Hopkins University Press, 1992.

Dairy Barns

BUCKENDORF, MADELINE. "Early Dairy Barns of Buhl." In Louie Attebury and Wayland Hand, eds., *Idaho Folklore*. Salt Lake City: University of Utah Press, 1985.

DURAND, LOYAL, JR. "Dairy Barns of South-eastern Wisconsin." *Economic Geography*, vol. 17 (1943).

Domestic Tankhouses

BOUCHER, AARON S., and ROBERT B. KENT. "Tankhouses in Nebraska: Distribution, Construction Styles, and Use." *Material Culture*, vol. 24, no. 1 (Spring 1992).

HARPER, GLEN A. "Historic Context for the Presence and Absence of Tankhouses in Ohio." *Material Culture*, vol. 24, no. 1 (Spring 1992).

KENT, ROBERT B. "Tankhouses on the High Plains of Western Kansas and Eastern Colorado." *Material Culture*, vol. 24, no. 1 (Spring 1992).

PITMAN, LEON S. "The Domestic Tankhouse as Vernacular Architecture in Rural California." *Material Culture*, vol. 24, no. 1 (Spring 1992).

PITMAN, LEON S. "Domestic Tankhouses of Rural California." *Pioneer America*, vol. 8, no. 2 (July 1976).

THORSHEIM, KATIE. "Running Water in Paradise: The Adoption and Diffusion of Domestic Tankhouses in Southern Oregon's Rogue Valley." *Material Culture*, vol. 24, no. 1 (Spring 1992).

Dutch Barns

FITCHEN, JOHN. *The New World Dutch Barn*. Syracuse: Syracuse University Press, 1968.

PRUDON, THEODORE H. M. "The Dutch Barn in America: Survival of a Medieval Structural Frame." *New York Folklife,* vol. 2 (Winter 1976).

WACKER, PETER. "Folk Architecture as an Indicator of Culture Areas and Culture Diffusion: Dutch Barns and Barracks in New Jersey." *Pioneer America,* vol. 5, no. 2 (July 1973).

English Barns

COFFEY, BRIAN. "Nineteenth-Century Barns of Geauga County, Ohio." *Pioneer America,* vol. 10, no. 2 (December 1978).

FINK, DANIEL. *Barns of the Genesee Country, 1790–1915.* Geneseo, N.Y.: James Brunner Publishers, 1988.

NOBLE, ALLEN G., and RICHARD K. CLEEK. "Sorting Out the Nomenclature of English Barns." *Material Culture,* vol. 26, no. 1 (1994).

Fences

KILPINEN, JON T. "Traditional Fence Types of Western North America." *P.A.S.T.: Pioneer America Society Transactions,* vol. 15 (1992).

LEECHMAN, DOUGLAS. "Good Fences Make Good Neighbors." *Canadian Geographic Journal,* vol. 47 (1953).

MATHER, EUGENE COTTON, and JOHN FRASER HART. "Fences and Farms." *Economic Geography,* vol. 44, no. 2 (1954).

MURRAY-WOOLEY, CAROLYN, and KARL RAITZ. *Rock Fences of the Bluegrass.* Lexington: University Press of Kentucky, 1992.

NORRIS, DARRELL A. "Ontario Fences and the American Scene." *American Review of Canadian Studies,* vol. 12, no. 2 (Summer 1982).

RIKOON, J. SANFORD. "Traditional Fence Patterns in Owyhee County, Idaho." *P.A.S.T.: Pioneer America Society Transactions,* vol. 7 (1984).

Finnish Dairy Barns

APPS, JERRY, and ALLEN STRANG. *Barns of Wisconsin.* Madison, Wisc.: Tamarack Press, 1977.

Finnish Hay Barns

KAUPS, MATTI. "Finnish Meadow-Hay Barns in the Lake Superior Region." *Journal of Cultural Geography,* vol. 10, no. 1 (Fall/Winter 1989).

German Barns (see also Pennsylvania Barns)

WILHELM, HUBERT G. H. "Double Overhang Barns in Southeastern Ohio." *P.A.S.T: Pioneer America Society Transactions,* vol. 12 (1989).

———. "German Settlement and Folk Building Practices in the Hill Country of Texas." *Pioneer America,* vol. 3, no. 2 (July 1971).

Grundscheier

KEYSER, ALLAN G., and WILLIAM P. STEIN. "The Pennsylvania German Tri-Level Ground Barn." *Der Reggeboge,* vol. 9, nos. 3–4 (1975).

Hay Barracks

BLACKBURN, RODERIC, and SHIRLEY DUNN. "The Hay Barrack: A Dutch Favorite." *Dutch Barn Preservation Society Newsletter*, vol. 2, no. 2 (Fall 1989).

McTERNAN, DON. "The Barrack: A Relict Feature on the North American Cultural Landscape." *P.A.S.T.: Pioneer America Society Transactions*, vol. 1 (1978).

NOBLE, ALLEN G. "The Hay Barrack: Form and Function of a Relict Landscape Feature." *Journal of Cultural Geography*, vol. 5, no. 2 (Spring/Summer 1985).

Hay Derricks

FIFE, AUSTIN E., and JAMES M. FIFE. "Hay Derricks of the Great Basin and Upper Snake River Valley." *Western Folklore*, vol. 7 (1948).

FRANCAVIGLIA, RICHARD V. "Western Hay Derricks: Cultural Geography and Folklore as Revealed by Vanishing American Technology." *Journal of Popular Culture*, vol. 11, no. 4 (Spring 1978).

Hay Hoods

FRANCAVIGLIA, RICHARD V. "Western American Barns: Architectural Form and Climatic Considerations." *Yearbook of the Association of Pacific Coast Geographers*, vol. 34 (1972).

Hay Stackers

ALWIN, JOHN A. "Montana's Beaverslide Hay Stacker." *Journal of Cultural Geography*, vol. 3, no. 1 (Fall/Winter 1982).

Hop Barns

CALKINS, CHARLES, and WILLIAM LAATSCH. "The Hop Houses of Waukesha County, Wisconsin." *Pioneer America*, vol. 9, no. 2 (December 1977).

DARLINGTON, JAMES. "Hops and Hop Houses in Upstate New York." *Material Culture*, vol. 16, no. 1 (Spring 1984).

NELSON, HERBERT B. "The Vanishing Hop-Driers of the Willamette Valley." *Oregon Historical Quarterly*, vol. 64 (September 1963).

Madawaska Twin Barns

KONRAD, VICTOR A., and MICHAEL CHANEY. "Madawaska Twin Barns." *Journal of Cultural Geography*, vol. 3, no. 1 (Fall/Winter 1982).

Mormon Cowsheds

LEE, DAVID R., and HECTOR H. LEE. "Thatched Cowsheds of the Mormon Country." *Western Folklore*, vol. 40, no. 2 (April 1981).

New England Connected Barn

HUBKA, THOMAS C. *Big House, Little House, Back House, Barn: The Connected*

Farm Buildings of New England (Hanover, N.H.: University Press of New England, 1984).

———. "The Connected Farm Building of Southwestern Maine." *Pioneer America*, vol. 9, no. 2 (December 1977).

ZELINSKY, WILBUR. "The New England Connecting Barn." *Geographical Review*, vol. 48, no. 3 (October 1958).

Owl Holes

MCILWRAITH, THOMAS F. "The Diamond Cross: An Enigmatic Sign in the Rural Ontario Landscape." *Pioneer America*, vol. 13, no. 1 (March 1981).

WILHELM, HUBERT G. H. "Owl Holes: A Settlement Residual in Southeastern Ohio." *Ohio Geographers: Recent Research Themes*, vol. 16 (1986).

Pennsylvania Barns (see also German Barns)

BASTIAN, ROBERT W. "Southeastern Pennsylvania and Central Wisconsin Barns: Examples of Independent Parallel Development?" *Professional Geographer*, vol. 27, no. 2 (May 1975).

CALKINS, CHARLES F., and MARTIN C. PERKINS. "The Pomeranian Stable of Southeastern Wisconsin." *Concordia Historical Institute Quarterly*, vol. 53, no. 3 (1980).

DORNBUSH, CHARLES H., and J. K. HEYL. *Pennsylvania German Barns*, vol. 31. Allentown, Pa.: Pennsylvania German Folklore Society, 1965.

ENSMINGER, ROBERT F. *The Pennsylvania Barn: Its Origin, Evolution, and Distribution in North America*. Baltimore: Johns Hopkins University Press, 1992.

———. "A Search for the Origin of the Pennsylvania Barn." *Pennsylvannia Folklife*, vol. 30, no. 2 (Winter 1980–81).

GLASS, JOSEPH W. *The Pennsylvania Culture Region: A View from the Barn*. Ann Arbor, Mich.: UMI Research Press, 1986.

GLASSIE, HENRY. "The Pennsylvania Barn in the South, Part 1." *Pennsylvania Folklife*, vol. 15, no. 2 (Winter 1965–66).

———. "The Pennsylvania Barn in the South, Part 2." *Pennsylvania Folklife*, vol. 15, no. 4 (Summer 1966).

JORDAN, TERRY G. "Alpine, Alemannic and American Log Architecture." *Annals of the Association of American Geographers*, vol. 70, no. 2 (June 1980).

KAUFFMAN, HENRY J. "Pennsylvania Barns." *Farm Quarterly*, vol. 9, no. 3 (1954).

RIDLEN, SUSANNE S. "Bank Barns in Cass County, Indiana." *Pioneer America*, vol. 4, no. 2 (July 1972).

SHOEMAKER, ALFRED L. *The Pennsylvania Barn*. Lancaster, Pa.: Pennsylvania Dutch Folklore Center, 1955.

WILHELM, HUBERT G. H. "The Pennsylvania Dutch Barn in Southeastern Ohio." *Geoscience and Man*, vol. 5 (1974).

Rack-Side Barns

RAITZ, KARL B. "The Barns of Barren County." *Landscape*, vol. 22, no. 2 (1978).

Raised Barns

GLASSIE, HENRY. "The Variation of Concepts within Tradition: Barn Building in Otsego County, New York." *Geoscience and Man*, vol. 5 (1974).

NOBLE, ALLEN G., and RICHARD K. CLEEK. "Sorting Out the Nomenclature of English Barns." *Material Culture*, vol. 26, no. 1 (1994).

Roofs, Decorated

SELZ, SHARON. "Roofing Ex-Farmer's Really Right on Top of Things!" *Farm and Ranch Living*, vol. 12, no. 6 (February/March 1990).

STEPHENS, DAVID T., and ALEX T. BOBERSKY. "Dated Slate Roofs of Columbiana County, Ohio." *P.A.S.T.: Pioneer America Society Transactions*, vol. 6 (1983).

Round and Polygonal Barns

HANOU, JOHN T. *A Round Indiana: Round Barns in the Hoosier State*. West Lafayette, Ind.: Purdue University Press, 1993.

PRICE, H. WAYNE, and KEITH SCULLE. "The Failed Round Barn Experiment." *P.A.S.T.: Pioneer America Society Transactions*, vol. 6 (1983).

SOIKE, LOWELL J. *Without Right Angles: The Round Barns of Iowa*. Des Moines: Iowa State Historical Department, 1983.

Silos

FISH N. S. "The History of the Silo in Wisconsin." *Wisconsin Magazine of History*, vol. 8 (December 1924).

MILES, MANLY. *Silos, Ensilage and Silage*. New York: Orange Judd Co., 1913.

NOBLE, ALLEN G. "Barns and Square Silos in Northeastern Ohio." *Pioneer America*, vol. 6, no. 2 (July 1974).

———. "Diffusion of Farm Silos." *Landscape*, vol. 25, no. 1 (February 1981).

———. "The Evolution of American Farm Silos." *Journal of Cultural Geography*, vol. 1, no. 1 (1980).

———. "The Silo in the Eastern Midwest: Patterns of Evolution and Distribution." *Ohio Geographers: Recent Research Themes*, vol. 4 (1976).

SUTER, ROBERT. *The Courage to Change*. Danville, Ill.: Interstate Printers and Publishers, 1964.

Smokehouses

LONG, AMOS, JR. "Smoke Houses in the Lebanon Valley." *Pennsylvania Folklife*, vol. 13 (Fall 1962).

REHDER, JOHN B., JOHN MORGAN, and JOY L. MEDFORD. "The Decline of Smokehouses in Grainger County, Tennessee." *West Georgia College Studies in the Social Sciences*, vol. 18 (June 1979).

WACKER, PETER O. "Cultural and Commercial Regional Associations of Traditional Smoke-houses in New Jersey." *Pioneer America*, vol. 3, no. 2 (July 1971).

Stone Fenceposts

RAFFERTY, MILTON D. "The Limestone Fenceposts of the Smokey Hill Region of Kansas." *Pioneer America,* vol. 6, no. 1 (January 1974).

Stone Walls

STRAIGHT, STEPHEN. "Stone Walls." *P.A.S.T.: Pioneer America Society Transactions,* vol. 10 (1987).

MURRAY-WOOLEY, CAROLYN, and KARL RAITZ. *Rock Fences of the Bluegrass.* Lexington: University Press of Kentucky, 1992.

Stovewood Architecture

LEDOHOWSKI, EDWARD, and DAVID BUTTERFIELD. In *Architectural Heritage: The Eastern Interlake Planning Districts,* 92–95. Winnipeg: Manitoba Department of Cultural Affairs and Historical Resources, 1983.

PERRIN, RICHARD. "Wisconsin's Stovewood Architecture." *Wisconsin Academy Review,* vol. 20, no. 2 (1974).

TISHLER, WILLIAM. "Stovewood Architecture." *Landscape,* vol. 23, no. 3 (1979).

———. "Stovewood Construction in the Upper Midwest and Canada: A Regional Vernacular Architectural Tradition." In Camille Wells, ed., *Perspectives in Vernacular Architecture.* Annapolis, Md.: Vernacular Architecture Forum, 1982.

Springhouses

GLASSIE, HENRY. "The Smaller Outbuildings of the Southern Mountains." *Mountain Life and Work,* vol. 40 (1964).

Summer Kitchens

KURTI, LASZLO. "Hungarian Settlement and Building Practices in Pennsylvania and Hungary: A Brief Comparison." *Pioneer America,* vol. 12, no. 1 (February 1980).

LANGLEY, THORPE. "Geography of the Maple Area, Douglas County, Wisconsin." M.A. thesis, University of Wisconsin, Madison, 1932.

LESSARD, MICHELLE, and GILLES VILANDRE. *La maison traditionelle au Québec.* Montreal: Les Editions de l'Homme, 1972.

LONG, AMOS, JR. "Pennsylvania Summer Houses and Summer Kitchens." *Pennsylvania Folklife,* vol. 15 (1965).

WILLIAMS, HATTIE. "A Social Study of the Russian German." *University Studies of the University of Nebraska,* vol. 16, no. 3 (1916).

Swedish Barns

A:SON-PALMQVIST, LENA. *Building Traditions among Swedish Settlers in Rural Minnesota.* Stockholm: Nordiska Museet/The Emigrant Institute, 1983.

Tobacco Barns

FLYNN, LIGON, and ROMAN STANKUS. "Carolina Tobacco Barns: Form and Significance." In Doug Swaim, ed., *Carolina Dwelling*. (Raleigh: North Carolina State University, School of Design, 1978.

HART, JOHN FRASER, and EUGENE COTTON MATHER. "The Character of Tobacco Barns and Their Role in the Tobacco Economy of the United States." *Annals of the Association of American Geographers,* vol. 51, no. 3 (September 1961).

RAITZ, KARL B. "The Wisconsin Tobacco Shed: A Key to Ethnic Settlement and Diffusion." *Landscape*, vol. 20, no. 1 (October 1975).

SCISM, LAURA. "Carolina Tobacco Barn: History and Function." In Doug Swaim, ed., *Carolina Dwelling*. Raleigh: North Carolina State University, School of Design, 1978.

VOGELER, INGOLF, and THOMAS DOCKENDORFF. "Central Minnesota Relic Tobacco Shed Region." *Pioneer America*, vol. 10, no. 2 (December 1978).

WILHELM, HUBERT. "Tobacco Barns and Pent Roofs in Western Ohio." *P.A.S.T.: Pioneer America Society Transactions*, vol. 6 (1983).

Transverse Frame Barns

SCULLE, KEITH A., and H. WAYNE PRICE. "The Traditional Barns of Hardin County, Illinois: A Survey and Interpretation." *Material Culture*, vol. 25, no. 1 (Spring 1993).

Welsh Barns

BROWN, MARY ANN. "Barns in the Black Swamp of Northwestern Ohio." *P.A.S.T.: Pioneer America Society Transactions*, vol. 12 (1989).

Windmills

BAKER, T. LINDSAY. *A Field Guide to American Windmills*. Norman: University of Oklahoma Press, 1985.

NOBLE, ALLEN G. "Windmills in American Agriculture." *Material Culture*, vol. 24, no. 1 (Spring 1992).

Wisconsin Dairy Barns

BUCKENDORF, MADELINE. "Early Dairy Barns of Buhl." In Louie Attebury and Wayland Hand, eds., *Idaho Folklife*. Salt Lake City: University of Utah Press, 1985.

Barn Report Form

Your name:_____ Date:_____

Street address:_____

City/state/zip:_____ Telephone:_____

Location of Barn

State or province:_____

Township and/or county:_____

Road location:_____

_____

Type of barn:_____

Roof type:_____ Floor plan:_____

Dimensions:_____

Door placement:_____

Foundation material:_____

Wall material:_____

Ventilation:_____

Decorative details:_____

Other comments:_____

_____

_____

Are photos attached?   YES   NO

Did you report the barn

    because it is out of its reported range?   YES   NO

    because it might be a new type?   YES   NO

    other, please comment:_____

_____

Were there other barns like this in the area?   YES   NO

    If yes, how many and how widespread?_____

_____

Please sketch the barn on the back of this page.

Mail to: North American Barn Project
        University of Wisconsin
        400 University Drive
        West Bend, WI 53095

Index

ABOUT THE AUTHORS

ALLEN G. NOBLE is Professor of Geography at the University of Akron and is Director Emeritus of the Pioneer America Society. He is the author of *Wood, Brick, and Stone: The North American Settlement Landscape,* for which he won the Honors Award of the Association of American Geographers, and author or editor of books and articles on the vernacular architecture of North America.

RICHARD K. CLEEK is Chief Information Officer for the University of Wisconsin Colleges. His research and publications have been in the fields of medical geography and geographic education. He is currently investigating the large variety of barn types and their geographic patterns on the Wisconsin landscape.